THERMAL PROPERTIES OF GREEN
AND BIOCOMPOSITES

W9-CRB-798

Hot Topics in Thermal Analysis and Calorimetry

Volume 4

Series Editor:
Judit Simon, *Budapest University of Technology and Economics, Hungary*

Thermal Properties of Green Polymers and Biocomposites

by

Tatsuko Hatakeyama

*Otsuma Women's University,
Tokyo, Japan*

and

Hyoe Hatakeyama

*Fukui University of Technology,
Japan*

KLUWER ACADEMIC PUBLISHERS
DORDRECHT / BOSTON / LONDON

Chemistry Library

A C.I.P. Catalogue record for this book is available from the Library of Congress.

ISBN 1-4020-1907-6 (HB)
ISBN 1-4020-2354-5 (e-book)

Published by Kluwer Academic Publishers,
P.O. Box 17, 3300 AA Dordrecht, The Netherlands.

Sold and distributed in North, Central and South America
by Kluwer Academic Publishers,
101 Philip Drive, Norwell, MA 02061, U.S.A.

In all other countries, sold and distributed
by Kluwer Academic Publishers,
P.O. Box 322, 3300 AH Dordrecht, The Netherlands.

Printed on acid-free paper

Printed in the Netherlands.

Contents

Preface vii

List of Abbreviations ix

Chapter 1
INTRODUCTION
1. Overview of Green Polymers 1
2. Molecular Level Morphology of Important Green Polymers: Cellulose and Lignin 3
3. Raw Materials for Synthetic Green Polymers: Molasses and Lignin ... 7
4. Scope of This Book 9

Chapter 2
CHARACTERIZATION OF GREEN POLYMERS
1. Thermal Analysis 13
2. Other Characterization Methods 25

Chapter 3
THERMAL PROPERTIES OF CELLULOSE AND ITS DERIVATIVES
1. Introduction 39
2. Thermal Properties of Cellulose in Dry State 42
3. Cellulose-Water Interaction 56
4. Liquid Crystals and Complexes 84
5. Hydrogels 108
6. Thermal Decomposition of Cellulose and Related Compounds 116

Chapter 4
Polysaccharides from plants
   1. Gelation 131
   2. Glass Transition and Liquid Crystal Transition.............................. 155

Chapter 5
Lignin
   1. Introduction 171
   2. Glass Transition of Lignin in Solid State 173
   3. Heat Capacity and Enthalpy Relaxation of Lignin.................... 184
   4. Molecular Relaxation 188
   5. Lignin-Water Interaction 198
   6. Thermal Decomposition 208

Chapter 6
PCL DERIVATIVES FROM SACCHARIDES, CELLULOSE AND LIGNIN
   1. Polycaprolactone Derivatives from Saccharides and Cellulose 217
   2. Polycaprolactone Derivatives from Lignin.......................... 238

Chapter 7
ENVIRONMENTALLY COMPATIBLE POLYURETHANES DERIVED FROM SACCHARIDES, POLYSACCHARIDS AND LIGNIN
   1. Polyurethane Derivatives from Saccharides..................... 249
   2. Polyurethanes Derived from Lignin 273
   3. Saccharides- and Lignin-Based Hybrid Polyurethane Foams 293

Chapter 8
BIO- AND GEO-COMPOSITES CONTAINING PLANT MATERIALS
   1. Biocomposites Containing Cellulose Powder and Wood Meal........ 305
   2. Biocomposites Containing Coffee Grounds 309
   3. Geocomposites 314

Subject Index... 325

Preface

In recent years, green polymers have received particular attention, since people have become more environmentally conscious. During the last fifty years, green polymers have sometimes been neglected compared to more high profile research subjects in academic and industrial fields. The authors of this book have continuously made efforts to investigate the properties, especially thermal properties, of green polymers and to extend their practical applications. Hence, the first half of this book is devoted to our results on fundamental research and the second half describes our recent research, mainly based on the authors' patents.

The authors are grateful to our long term friends; Professor Clive Langham, Nihon University, to whom we are especially grateful for his editorial advice, Professor Kunio Nakamura, Otsuma Women's University, Dr. Shigeo Hirose, National Institute of Advanced Science and Technology, Professor Shoichiro Yano, Nihon University, Professor Hirohisa Yoshida, Tokyo Metropolitan University, Dr. Francis Quinn, Loreal Co., Professor Masato Takahashi, Shinshu University, Dr. Per Zetterlund, Kobe University, and Dr. Mika Iijima, Yokkaichi University. We also wish to thank Ms. Chika Yamada for her helpful assistance.

As Lao Tse, the ancient Chinese philosopher said, "materials that look fragile and flexible, like water, are the original matters of the universe". The authors hope that green polymers on the earth continue to coexist with us in the long term incarnation of the universe.

Hyoe Hatakeyama
Tatsuko Hatakeyama

List of abbreviations

AFM   atomic force microscopy
AL   alcoholysis lignin (Alcel lignin)
ALPCL   alcoholysis lignin-based PCL
CA   cellulose acetate
CAPCL   cellulose acetate-based PCL
CG   coffee ground
CL   ε-caprolactone
CMC   carboxymethylcellulose
CellPCL cellulose-based polycaprolactone derivatives
C_p heat capacity
DABCO   1,4-diazobicyclo(2,2,2)octane
DBTDL   di-n-butyltin dilaurate
DEG   diethylene glycol
DMA   dynamic mechanical analysis
DMAc   N, N-dimethylacetoamide
DPPH   1,1-diphenyl-2-picrylhydrazyl
DS   degree of substitution
DSC   differential scanning calorimetry
DTA   differential thermal analysis
DTA-TG   differential thermal analysis-thermogravimetry
DTG   derivative thermogravimetry
DT_d derivative thermal decomposition temperature
ESR   electron spin resonance
E_a activation energy
E' dynamic storage modulus

E''	dynamic loss modulus
FTIR	Fourier transform infrared spectrometry
Fru	fructose
GP	graft polyol (styrene- and acrylonitrile grafted polyether)
Glu	glucose
KL	kraft lignin
KLDPU	kraft lignin-based diethylene glycol type polyurethane
KLPCL	kraft lignin-based PCL
KLPPU	kraft lignin-based polyethylene glycol type polyurethane
KLTPU	kraft lignin-based triethylene glycol type polyurethane
LDI	lysine diisocyante
LS	lignosulfonate
LSDPU	lignosulfonate-based diethylene glycol type polyurethane
LSPCL	lignosulfonate-based polycaprolactone
LSPPU	lignosulfonate-based polyethylene glycol type polyurethane
LSTPU	lignosulfonate-based triethylene glycol type polyurethane
LTI	lysine triisocyante
LiCL	lithium chloride
Lig	lignin
LigPCL	lignin-based PCL
M	mass
MDI	diphenylmethane diisocyanate [poly (phenylene methylene) polyisocyanate
ML	molasses
MLP	molasses polyol
MR	mass residue
MWL	milled wood lignin
NCO/OH	isocyanate group/hydroxyl group ratio
NMR	nuclear magnetic resonance spectrometry
NaCS	Sodium cellulose sulfate
OHV	hydroxyl group value
PCL	polycaprolactone
PEG	polyethylene glycol
PEP	polyester polyol
PPG	poly(propylene glycol)
PSt	polystyrene
PU	polyurethane
PVA	poly(vinyl alcohol)
PVP	poly(vinyl pyroridone)
RH	relative humidity, %
SEM	scanning electron microscopy
Suc	sucrose

T	temperature
TBA	torsion braid analysis
TDI	tolylene diisocyanate
TEG	triethylene glycol
TG	thermogravimetry
TMA	thermomechanical analysis
TMAEP	trimethylaminoethylpiperazine
T_{cc}	cold-crystallization temperature
T_d	thermal degradation temperature
T_g	glass transition temperature
T_m	melting temperature
WAX	wide line x-ray diffractometry
W_c	water content= mass of water / mass of dry sample, $g\ g^{-1}$
$\tan \delta$	$=E''/E'$
ΔC_p	heat capacity difference at T_g
ΔH_m	enthalpy of melting
ε	strain
ρ	apparent density
σ	strength

Chapter 1

INTRODUCTION

1. OVERVIEW OF GREEN POLYMERS

Synthetic polymers are essential for modern human life, since they are used in industrial and agricultural fields. However, most synthetic polymers that have been developed by using petroleum and coal as raw materials are not compatible with the environment, since they cannot be included in the natural recycling system. There are serious contradictions between the convenience that people require today and compatibility with the natural environment. It is easy to say that we should use only natural materials in order to solve the problems coming from man-made materials. However, this means that we lose all the convenient features and materials which science has developed through human history. Therefore, development of environmentally compatible polymers (**green polymers**) is the key to sustainable developments that can maintain our rich and convenient life. Table 1-1 offers an overview of green polymers that have recently been developed.

In order to develop green polymers, it is essential to understand that nature constructs a variety of materials that can be used. Saccharides have already been used extensively in the food, medical and cosmetic industries. Plant materials such as cellulose, hemicellulose and lignin are the largest organic resources. However, it can be said that the above natural polymers, except for cellulose, are not very well used. Hemicellulose has not yet been utilized. Lignin, which is obtained as a by-product of the pulping industry is mostly burnt as fuel and only increases the amount of carbon dioxide in the environment, although lignin is one of the most useful natural resources.

Table 1-1. Examples of green polymers

Polymer types	Examples
From microorganisms	Polysaccharides such as xanthan gum, alginic acid, hyaluronan, and gellan gum
	Polyesters such as poly(hydroxyalkanoate)s
From plants	Polysaccharides such as cellulose, lignin, starch, carrageenan, and locust bean gum
	Cellulose esters such as cellulose acetates
	Saccharide-based polyurethanes and polycaprolactone derivatives
	Lignin-based polyurethanes and poly-caprolactone derivatives
	Starch-based blends
From animals	Collagen, Chitin
	Chitin and chitosan-based polymeric derivatives and composites

Biomaterials span the range from elastic solids to viscous liquids. However, they have been difficult to use as natural resources for polymers that are useful for human life because of the complexity based on the intricacies of their molecular architecture. However, scientific advances enable us to understand molecular features of biomaterials through modern analytical methods. Now it is the time to consider that the compounds produced through biosynthesis can be used as half-made up raw materials for the synthesis of useful plastics and materials. Major plant components, such as saccharides and lignin, contain highly reactive hydroxyl groups that can be used as reactive chemical reaction sites. As shown in Figure1-1, it is possible to convert saccharides and lignin to various green polymers that are environmentally compatible [1-26].

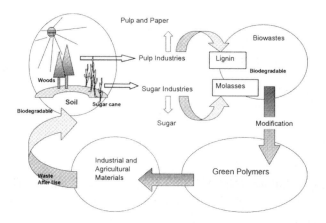

Figure 1-1. Circulation of lignin- and saccharide-based synthetic polymers in nature.

This book is concerned with the thermal properties of green polymers such as natural polymers and polymers derived from saccharides and lignins. The above green polymers include polymers such as poly(ε-caprolactone) (PCL) and polyurethane (PU) derivatives. PCL derivatives were synthesized from lignin, saccharides, cellulose and cellulose acetates. PU derivatives were prepared from saccharides and lignins. Thermal properties of the above polymers were characterized by various thermal analyses including thermogravimetry (TG), differential thermal analysis (DTA), differential scanning calorimetry (DSC), thermomechanometry (TMA) and dynamic mechanical analysis (DMA). Simultaneous measurements combining various techniques such as TG-Fourier transform-infrared spectrometry (FTIR) and TG-DTA are also mentioned.

2. MOLECULAR LEVEL MORPHOLOGY OF IMPORTANT GREEN POLYMERS: CELLULOSE AND LIGNIN

The molecular architecture of cellulose and lignin has received particular attention for over 100 years, since both biopolymers are the major components of plant materials. Due to recent studies performed by x-ray diffractometry and solid state nuclear magnetic resonance spectrometry (NMR), the crystalline structure of cellulose has been investigated. In contrast, the higher-order structure of lignin in the amorphous state has scarcely been studied, since analytical methods were limited. The results were averaged over the number of molecules based on indirect analysis. Recently, the supermolecular structure of biopolymers has been investigaited in nano-level, since it is possible to observe individual molecules and molecular assemblies by atomic force microscopy (AFM) [27]. AFM directly visualizes the heterogeneity of biopolymers either in crystalline or amorphous state. Furthermore, morphological observation can be correlated with the results obtained by other physical measurements.

AFM has been used in order to observe the supermolecular structure of cellulose and lignin by using their water soluble derivatives such as sodium carboxymethylcellulose (NaCMC), sodium cellulose sulfate (NaCS) and sodium lignosulfonate (LS). Water soluble derivatives were used as samples, since aqueous solutions of samples were easy to spread on a freshly cleaved mica surface. The samples spread on mica were imaged by AFM.

An AFM image of NaCMC is shown in Figure 1-2. Rigid strands are clearly observed. The thickness of strands is ca. 0.7 nm, which strongly indicates that NaCMC molecules extended on mica surface are in mono- or double layers. It is considered that the hydrophobic side of molecules attaches to the mica surface and the carboxymethyl groups extend to the outer surface. The width of the strands ranges from 15.2 to 18.2 nm. When the results obtained by x-ray diffractometry are taken into consideration, 4 to 5 molecules are bundled and observed as a strand. In the above calculation, the size of the geometrical shape of the needle and the samples are calibrated. Figure 1-3 shows a three dimensional AFM image of NaCMC.

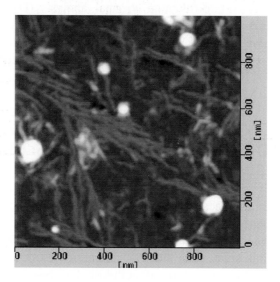

Figure 1-2. AFM image of sodium carboxymethylcellulose (NaCMC, concentration 10 μg ml^{-1}) showing extended molecular chain.

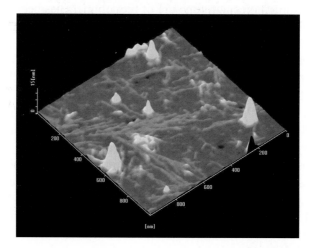

Figure 1-3. Three dimensional AFM image of NACMC (concentration 10 μg ml^{-1}).

Figures 1-4 and 1-5 show two and three dimensional AFM images of NaCS. Both figures indicate that sodium cellulose sulfate (NaCS) molecules show worm-like structures. The difference of the molecular shape between NaCMC and NaCS may be caused by the difference of substituted groups and also the degree of substitution (DS).

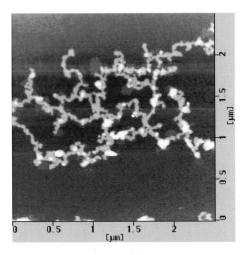

Figure 1-4. AFM image of sodium cellulose sulfate (NaCS) showing worm-like molecular chain structure (concentration 10 μg ml^{-1}).

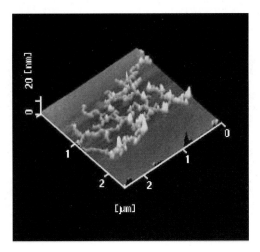

Figure 1-5. Three dimensional AFM image of NaCS showing worm-like molecular chain structure (concentration 10 μg ml^{-1}).

Figures 1-6 and 1-7 show two and three dimensional AFM images of sodium lignosulfonate (NaLS). Both figures show that lignin has a complicated network structure that is highly crosslinked.

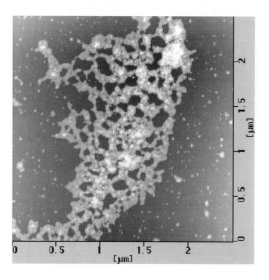

Figure 1-6. AFM image of sodium lignosulfonate (NaLS) showing molecular chain forming network structure (concentration 10 μg ml⁻¹).

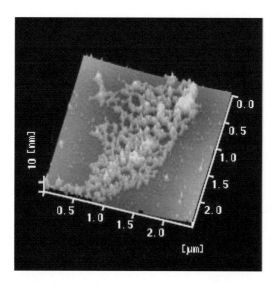

Figure 1-7. Three dimensional AFM image of sodium lignosulfonate (NaLS) showing molecular chain forming network structure (concentration 10 μg ml⁻¹).

3. RAW MATERIALS FOR SYNTHETIC GREEN POLYMERS: MOLASSES AND LIGNIN

3.1 Molasses

Molasses is a brown viscous liquid and is produced from sugar cane and beet. The chemical components of molasses consist of sucrose and saccharides such as glucose and fructose. An example of the chemical components of molasses is shown in Table 1-2. Molasses is usually used as an ingredient in the fermentation industry and also for livestock feed. However, it has been found that it is useful as a raw material for the synthesis of saccharide-based polyurethanes and polycaprolactones [1-11]. Molasses from sugar cane is produced in tropical and subtropical regions such as Brazil, Cuba, Thailand, Indonesia, Philippines and Okinawa.

Table 1-2. Chemical components of molasses [26]

Chemical components	Content / %
Sucrose	32.5
Glucose	8.5
Fructose	9.2
Other Saccharides	2.3
Water	20.5
Ash	9.5

On the other hand, beet molasses is produced in cold regions such as northern Europe, Russia and Hokkaido, Japan. Recent sugar production in the world is ca. 130 million tons / year. Production of molasses corresponds to ca. 30 % of sugar production. Accordingly, it is considered that 40 million tons / year of molasses is produced in the world. This amount seems to be more than enough for the production of environmentally compatible bio-based polymers in the future.

3.2 Lignins

Lignins are derived from renewable resources such as trees, plants, and agricultural crops. About 30 % of wood constituents are lignin. Lignins are nontoxic and extremely versatile in performance. Most industrial lignins are obtained from kraft and sulfite pulping processes. Kraft lignin is usually burnt as fuel at pulping mills. Annual lignin production in Japan is estimated to be about 8 million tons. Lignin production in the world is approximately 30 million tons / year. However, it should be noted that this value is only an estimate, since there are no reliable statistics on lignin production because it is mostly burnt as a fuel immediately after production. About one million tons of water soluble lignosulfonate derivatives which are by-products of sulfite pulping are consumed in Japan as chemicals such as dispersants [28].

Commercial lignin is a by-product of the pulping industry, as mentioned above, and is separated mostly from wood by a chemical pulping process. As described above, major delignification technolgies used in the pulping process are kraft and sulfite methods. Other delignification technologies are solvolysis processes using organic solvents or high pressure steam treatments to remove lignins from plants.

Since lignins are natural polymers with random crosslinkings, their physical and chemical properties differ depending on extraction processes. A part of the schematic chemical structure of lignin is shown in Figure 1-9.

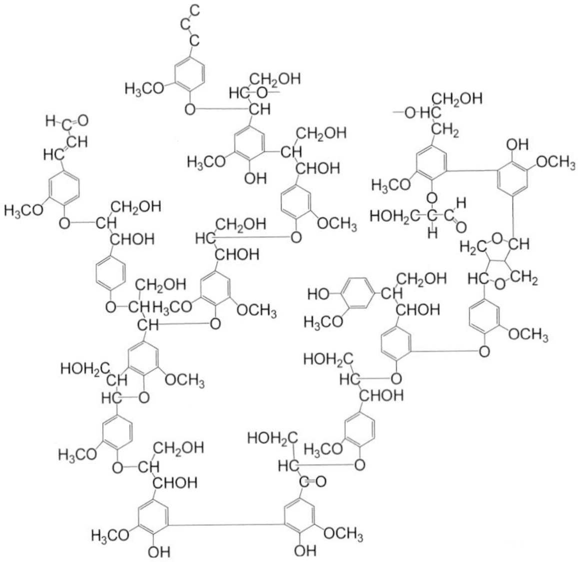

Figure 1-8. A part of schematic chemical structure of lignin [29].

As described in Chapter 5 of this book, the higher-order structure of lignin, which consists of phenyl propane units, is fundamentally amorphous. Three phenylpropaniod monomers such as coniferyl alcohol, synapyl alcohol and *p*-coumaryl alcohol are conjugated to produce a three dimensional lignin polymer in the process of radical-based lignin biosynthesis. For the above reason, lignin does not have a regular structure like cellulose, but is a physically and chemically heterogeneous material, although the exact chemical structure is unknown.

Since each lignin molecule has more than two hydroxyl groups, lignin-based polyurethane derivatives, polycaprolactone derivatives and epoxy resins are obtainable by using the hydroxyl group as the reaction site [1,2, 12-26].

4. SCOPE OF THIS BOOK

This book is concerned with characterization of polymers such as cellulose, lignin and green polymers by thermal and mechanical analyses, spectroscopy, and x-ray diffractometry. Synthesis of green polymers derived from saccharides and lignins, such as polyurethane and polycaprolactone derivatives having saccharide and lignin structures in the molecular chain is also described.

This book consists of 8 chapters. In Chapter 1, " Introduction", the background and objectives of this book are introduced. Chapter 2 is concerned with various analytical methods that are useful for the characterization of green polymers. The analytical methods are thermal analyses, such as differential scanning calorimetry (DSC), thermogravimetry (TG) and TG-Fourier transform-infrared spectrometry (TG-FTIR), spectroscopy such as infrared spectroscopy and nuclear magnetic resonance spectroscopy (NMR), microscopy such as polarizing microscopy, scanning electron microscopy and atomic force microscopy, and x-ray diffractometry. Chapter 3 is devoted to the discussion of thermal properties of cellulose, cellulose-water interaction, liquid crystals from water-soluble cellulose derivatives and hydrogels. Chapter 4 is on hydrogels and liquid crystals of various polysaccharides. Chapter 5 concerns various properties of lignins. Chapter 6 is concerned with polycaprolactone derivatives having cellulose and lignin structures in the molecular chain. Chapter 7 deals with polyurethane derivatives from saccharides and lignin. Chapter 8 describes biocomposites containing plant and inorganic materials.

REFERENCES

1. Hatakeyama, H., 2002, Thermal analysis of environmentally compatible polymers containing plant components in the main chain. *J. Therm. Anal. Cal.*, **70**, 755-759.
2. Hatakeyama, H., Asano, Y. and Hatakeyama, T., 2003, Biobased polymeric materials. In *Biodegradable Polymers and Plastics* (Chellini, E. and Solario, R. eds.), Kluwer Academic / Plenum Publishers, New York, pp. 103-119.
3. Hirose, S. Kobashigawa K. and Hatakeyama, H. 1994, Preparation and physical properties o f polyurethanes derived from molasses. *Sen-i Gakkaishi*, **50**, 538-542.
4. Morohoshi, N., Hirose S., Hatakeyama, H., Tokashiki, T. and Teruya, K., 1995, Biodegradation of polyurethane foams derived from molasses. *Sen-i Gakkaishi*, **51**, 143-149.

5. Zetterlund, P., Hirose, S., Hatakeyama, T., Hatakeyama, H. and Albertsson, A-C., 1997, Thermal and mechanical properties of polyurethanes derived from mono- and disaccharides. *Polym. Inter.*, **42**, 1-8.

6. Hatakeyama, H., Kobahigawa, K., Hirose, S. and Hatakeyama, T., 1998, Synthesis and physical properties of polyurethanes from saccharide-based polycaprolactones. *Macromol. Symp.*, **130**, 127-138.

7. Hatakeyama, T., Tokashiki, T. and Hatakeyama, H., 1998, Thermal properties of polyurethanes derived from molasses before and after biodegradation, *Macromol. Symp.*, **130**, 139-150.

8. Hatakeyama, H., 2000, Adaptation of plant components in molecular of environmentally compatible polymers. *Petrotech*, **23**, 724-730.

9. Hatakeyama, H., 2001, Thermal properties of biodegradable polymers. *Netsu Sokutei*, **28**, 183-191.

10. Hatakeyama, H., 2001, Biodegradable polyurethane using saccharide and lignin. In *Practical Technology of Bio-degradable Plastics*, CMC, Tokyo, pp. 97-108.

11. Asano, Y., Hatakeyama, H., Hirose, S. and Hatakeyama, T., 2001, Preparation and physical properties of saccharide-based polyurethane foams. In *Recent Advances in Environmentally Compatible Polymers* (J. F. Kennedy, G. O. Philips, P. A. Williams and H. Hatakeyama eds.), Woodhead Publishing Ltd., Cambridge, UK, pp. 241-246.

12. Yoshida, H., Mörck, R., Kringstad, K. P. and Hatakeyama, H., 1990, Kraft lignin in polyurethanes. II. Effects of the molecular weight of kraft lignin on the properties of polyurethanes from a kraft lignin-polyether triol-polymeric MDI system. *J. Appl. Polym. Sci.*, **40**, 1819-1832.

13. Reimann, A., Mörck, R., Hirohisa, Y., Hatakeyama, H. and K. P. Kringstad, 1990, Kraft lignin in polyurethanes. III. Effects of the molecular weight of PEG on the properties of polyurethanes from a kraft lignin-PEG-MDI system. *J. Appl. Polym. Sci.*, **41**, 39-50.

14. Nakamura, K., Mörck, R., Reimann, A., Kringstad, K. P. and Hatakeyama, H., 1991, Mechanical properties of solvolysis lignin-derived polyurethanes. *Polymer for advanced technology*, **2**, 41-47.

15. Nakamura, K., Hatakeyama, T. and Hatakeyama, H., 1992, Thermal properties of solvolysis lignin-derived polyurethanes. *Polymer for advanced technology*, **3**, 151-155.

16. Hirose, S., Nakamura, K., Hatakeyama, H., Meadows, J., Williams, P. A. and Phillips, G. O., 1993, Preparation and mechanical properties of polyurethane foams from lignocellulose dissolved in polyethylene glycol. In *Cellulosics: Chemical, Biochemical and Materials* (J. F. Kennedy Williams P. A. and Phillips, G. O., eds.), Ellis Horwood Limited, Chichester, UK, pp. 317-331.

17. Nakamura, K., Hatakeyama, H., Meadows, J., Williams, P. A. and Phillips, G. O., 1993, Mechanical properties of polyurethane foams derived from eucalyptus kraft lignin, In *Cellulosics: Chemical, Biochemical and Materials* (J. F. Kennedy Williams P. A. and Phillips, G. O., eds.), Ellis Horwood Limited, Chichester, UK, pp. 333-340.

18. Hatakeyama, H., Hirose, S., Nakamura, K. and Hatakeyama, T. 1993, New types of polyurethanes derived from lignocellulose and saccharides, In *Cellulosics: Chemical, Biochemical and Materials* (J. F. Kennedy Williams P. A. and Phillips, G. O., eds.), Ellis Horwood Limited, Chichester, UK, pp. 525-536.

19. Hatakeyama, H., 1993, Molecular design of biodegradable plastics, *Kagaku to Seibutsu*, **31**, 308-311.

20. Hatakeyama, H., 1993, Biodegradable plastics derived from plant resources, *Mokuzai Kogyo*, **48**, 161-165.

21. Hatakeyama, H. and Hirose, S., 1994, Design of biodegradable materials. *Kogyo Zairyo*, **42**, 34-37.

22. Nakamura, K., Nishimura, Y., Hatakeyama, T. and Hatakeyama, H., 1995, Mechanical and thermal properties of biodegradable polyurethanes derived from sericin. *Sen-i Gakkaishi*, **51**, 111-117.

23. Tokashiki, T., Hirose, S. and Hatakeyama, H., 1995, Preparation and physical properties of polyurethanes from oligosaccharides and lignocellulose system. *Sen-i Gakkaishi*, **51**, 118-122.

24. Hirose, S., Kobashigawa, K. and Hatakeyama, T., 1996, Preparation and physical properties of biodegradable polyurethanes derived from the lignin-polyester-polyol system, In *Cellulosics: Chemical, Biochemical and Materials* (J. F. Kennedy Williams P. A. and Phillips, G. O., eds.), Ellis Horwood Limited, Chichester, UK, pp. 277-282.

25. Nakano, J., Izuta, Y., Orita, T., Hatakeyama, H., Kobashigawa, K., Teruya, K. and Hirose, S., 1997, Thermal and mechanical properties of polyurethanes derived from fractionated kraft lignin. *Sen-i Gakkaishi*, **53**, 416-422.

26. Hirose, S., Kobashigawa, K., Izuta, Y. and Hatakeyama, H., 1998, Thermal degradation of polyurethanes containing lignin structure by TG-FTIR. *Polymer International*, **47**, 1-8.

27. Bonnel, D., 2001, *Scanning Probe Microscopy and Spectroscopy, Second Edition*, Wiley-VCH, New York.

28. Machihara, A. and Kawamura, M., 2001, Recent utilization of lignin. In *Recent Advances in Technology for Wood Chemical* (G. Meshituka ed.), CMC, Tokyo, 127-137.

29. Lin, S. W. and Dence, C. W., 1992, Methods in Lignin Chemaistry, Spring Verlag, Berlin, pp. 3-16.

Chapter 2

CHARACTERIZATION OF GREEN POLYMERS

In this chapter, experimental techniques which are ordinarily used in investigation of green polymers and related compounds will briefly be introduced. Conformation of apparatuses, results and practical experimental conditions will be included. Apparatuses introduced here are commercially available and widely found in laboratories. Experimental conditions of thermal analysis are in a moderate temperature range in which green polymers are measurable.

1.   THERMAL ANALYSIS

Thermal analysis is defined as an analytical experimental technique which investigates the physical properties of a sample as a function of temperature or time under controlled conditions. This definition is broad and the following techniques are referred to conventionally as thermal analysis, i.e. thermogravimetry (TG), differential thermal analysis (DTA), differential scanning calorimetry (DSC), thermomechanometry (TMA) and dynamic mechanical analysis (DMA). Recently, simultaneous measurements combining various techniques are widely used In this section, TG-DTA, TG-Fourier transform infrared spectroscopy (TG-FTIR), DSC, TMA and DMA will briefly be introduced. Detailed information is shown elsewhere [1-36].

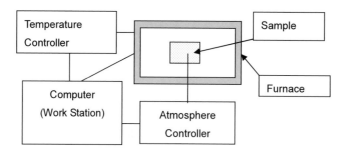

Figure 2-1. Conformation of thermal analysis apparatuses.

1.1   Thermogravimetry (TG)

1.1.1   TG-differential thermal analysis (DTA)

Thermogravimetry is the branch of thermal analysis which examines the mass change of a sample as a function of temperature in the scanning mode or as a function of time in the isothermal mode. A schematic conformation of a thermogravimeter is shown in Figure 2-2. At the present, almost all apparatuses used in the measurements of green polymers are those which enable simultaneous measurement of TG and differential thermal analysis (DTA) to be carried out. Balance systems, kinds of crucible, flow gas systems and other special attachments are described elsewhere in detail [6, 18, 32].

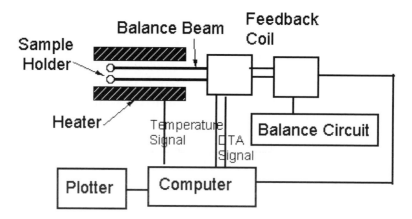

Figure 2-2. Schematic conformation of thermogravimeter.

In the investigation of green polymers, TG has been used in moderate conditions in order to obtain the following information.

1. Decomposition temperatures (T_{di}, T_d, T_{de}. etc)
2. Peak temperature of TG derivative curves (ΔT_{dp})
3. Mass residue at a temperature, range from 720 to 870 K (m_T)
4. Mass loss by vaporization of small molecular weight substances
5. Activation energy of decomposition and rate of decomposition

Standard TA computers are equipped with a software which determines the above basic results from (1) to (4). Additionally, a rate control program is commercially available [37, 38]. In order to measure green polymers, experimental conditions of TG which are ordinarily used in this book, are as follows; sample mass; 5 - 12 mg, material of crucible; platinum (carbon), shape of crucible: open and flat, temperature range; 290 - 870 K, heating rate (for standard measurements),10 - 20 K min^{-1}, heating rate (for calculation of kinetic parameters); 1 - 50 K min^{-1}, kinds of flow gas ; N_2, Air, or Ar (for special purpose), gas flow rate; 50 - 100 ml min^{-1}, respectively. Accuracy of data obtained by TG is found elsewhere [39]. Schematic TG curve and derivative curve are shown in Figure 2-3. T_d, ΔT_d, m_T are indicated using arrows. When two step decomposition is observed, the T_d is numbered from the low to high temperature side.

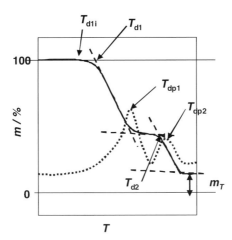

Figure 2-3. Schematic TG and TG derivative curves.

By using TG-FTIR, gases evolved from the sample decomposed in a TG sample cell are directly introduced to a FTIR sample cell and IR spectra are simultaneously measured as a function of temperature. In order to operate this apparatus properly, it is important to control the temperature of the transfer tube connecting TG with FTIR. Evolved gases condense in the tube if the temperature is low, at the same time, secondary decomposition takes place if the temperature is too high. Temperature and flow rate of purging gas of the connecting tube must be controlled appropriately. Various kinds of natural polymers have been measured by TG-FTIR, such as lignin [33, 40], polyurethane derived from saccharides [41] and polycaprolactone grafted cellulose acetate [42]. Based on the TG-FTIR data, the decomposition mechanism of green polymers has been investigated. Representative FTIR curves obtained by TG-FTIR are shown in Figure 2-4.

Experimental conditions for standard measurements of green polymers by TG-FTIR are as follows; sample mass; 5 -10 mg, heating rate; 10 or 20 K min^{-1}, gas flow rate; 100 ml min^{-1}, temperature range; 290 - 870 K. temperature of connecting tube; 520 K, resolution of FTIR; 1 cm^{-1} and acquisition time 10 scan sec^{-1}, respectively.

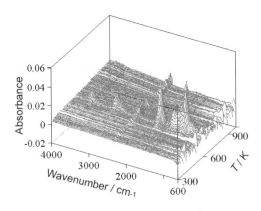

Figure 2-4. Three dimensional IR spectra as functions of wave numbers and temperature.

1.2   Differential scanning calorimetry (DSC)

Two types of DSC, power compensation type and heat flux type are used. In the power compensation type DSC, if a temperature difference is detected between the sample and reference, due to a phase change in the sample, energy is supplied until the temperature difference is less than a threshold value. In heat flux type DSC, the temperature difference between the sample and reference is measured as a function of temperature or time, under controlled temperature conditions. The temperature difference is proportional to the change in the heat flux.

When commercially available apparatuses of both types of DSC are compared, no large differences can be found concerning sensitivity, necessary amount of sample, temperature range of measurement, atmospheric gas supply, etc. Major differences between the two types of DSC are as follows; (1) due to the size of heater, isothermal measurements are easily carried out, when a power compensate type DSC is used. (2) due to the conformation of the sample cell, the low temperature measurements are carried out at a slow scanning rate, and a more stable baseline can be obtained by heat-flux type DSC.

Figure 2-5 shows a schematic conformation heat-flux type DSC and Figure 2-6 shows that of power compensation type DSC. Experimental conditions for standard measurements of green polymers by DSC are as follows; sample mass; 1 - 15 mg (ordinal condition, 5 - 7 mg), material of sample pan; Al (for solid and solution samples) and Ag (for dilute solution or hydrogels), shape of sample; open and flat type (for dry samples) and two different sealed types (for wet samples, solutions and hyrogels), temperature range; 120 K to a predetermined temperature lower than thermal decompositions (in standard conditions lower than 500 K), heating rate; 1 - 50 K min^{-1} (in standard conditions 10 K min^{-1}), atmospheric gas; N$_2$, gas flow rate; 30 ml min^{-1}. Repeatability and accuracy of DSC data of polymers are found elsewhere [43-45].

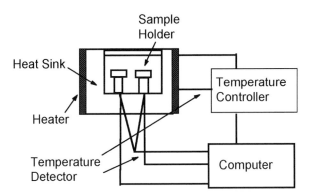

Figure 2-5. Schematic conformation of heat-flux type DSC.

By DSC, the following information on green polymers and related compounds is obtained.

1. The first order phase transition temperatures
2. Melting temperature (T_m)
3. Liquid crystal to liquid transition temperature (T_{lc-l})
4. Crystal to crystal transition

5. Crystallization temperature (T_c)
6. Cold crystallization temperature (T_{cc})
7. Pre-melt crystallization temperature (T_{pmc})
8. Liquid to liquid crystallization temperature (T_{l-lc})
9. Glass transition temperature (T_g)
10. Heat capacity difference at T_g (ΔC_p)

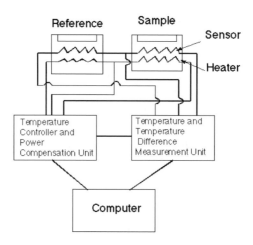

Figure 2-6. Schematic conformation of power compensation type DSC.

Figure 2-7 shows schematic DSC curves for the determination of transition temperatures and enthalpies. Ordinarily, peak temperature of melting (T_{pm}) and crystallization (T_{pc}) are used as an index of melting or crystallization temperature. It is noted that both temperatures are not obtained by equilibrium conditions. On this account, in this book the scanning rate is always shown in the figure captions. Scanning rate dependency of melting or crystallization of polymers is found elsewhere [29, 32].

Figure 2-8 shows a typical DSC heating curve of amorphous polymer. Glass transition is observed as a baseline deviation toward endothermic direction (direction of heat capacity increase). Due to the thermo-dynamically non-equilibrium nature of the glassy state, glass transition temperature (T_g) depends on the thermal history of a sample and measurement conditions such as the heating rate. On this account, the T_g value should always be stated along with precise experimental conditions and thermal history of the samples. In Figure 2-8, starting temperature (T_{ig}'), extrapolated temperature (T_{ig}), mid temperature (T_{mg}) and final temperature (T_{eg}) can be read. Generally T_{ig} or T_{mg} is reported as T_g. The above facts suggest that reported T_g values are not concrete values but depend on experimental conditions and definition of T_g.

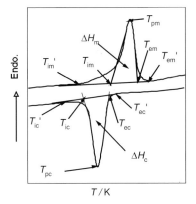

Figure 2-7. Schematic DSC curves for the determination of transition temperatures and enthalpies.

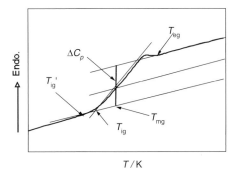

Figure 2-8. Schematic DSC heating curve showing glass transition. starting temperature (T_{gi}'), extrapolated temperature (T_{gi}), mid temperature (T_{gm}), final temperature (T_{gf}), heat capacity difference at T_{gi} (ΔC_p).

1.3   Thermomechanometry (TMA)

In thermomechanometry (thermomechanical analysis, TMA) the deformation of materials under constant stress, or constant strain, is measured as a function of temperature or time. Stress or strain can be applied to the sample in either a static or dynamic mode. Sample probes capable of measuring samples not only in air or inert gas but also in humid conditions or in water have been developed. A schematic conformation of a thermomechanometer is shown in Figure 2-9.

The following information can be obtained by static measurements of green polymers.

1. Glass transition temperature
2. Linear expansion or compression coefficient
3. Stress relaxation as a function of time at a predetermined temperature
4. Creep as a function of time at a predetermined temperature
5. Swelling rate and equilibrium swelling ratio under various stresses
6. Dynamic modulus, dynamic loss modulus and tan δ as a function of temperature.

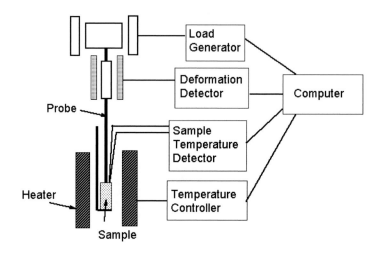

Figure 2-9. Schematic conformation of a thermomechanometer.

Softening temperature measured by TMA is practically used in commercial and industrial fields. Softening temperature is neither glass transition nor melting, but at a temperature higher than "softening temperature" thermoplastics start to flow. On this account, the softening temperature is an important index for polymer processing. Repeatability and reliability of TMA data is confirmed by a round robin test [46]. Almost all green polymers in the solid dry state lack flowability. On this account, in this book, softening temperature will not be described. Experimental conditions for standard measurements of green polymers by TMA are as follows; probe material; quartz, temperature range; 290 - 520 K (for dry sample), 273 - 263 K (for hydrogels). Applied stress, strain and frequencies have a wide range according to the kind of sample and shape of probe. Although there are various shapes of probe, two kinds of probe were used as shown in Figure 2-10.

Typical TMA curves in compression mode are shown in Figure 2-11. Transition temperature is determined as a cross point of two extrapolated lines as shown in the figure.

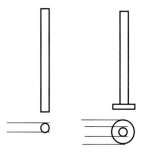

Figure 2-10. TMA probes used in the experiments shown in this book.

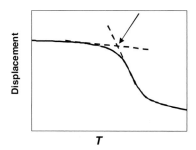

Figure 2-11. Schematic TMA curve in compression mode.

The sample holder for the measurement of swelling of samples is shown in Figure 2-12 [47]. The sample sheet was placed on a quartz plate and predetermined stress applied. Water is supplied from the bottom via a flexible tube. Deformation is detected as a function of time. When temperature dependency of swelling is measured, a water bath whose temperature is controllable was connected to the sample probe. Temperature was changed stepwise.

Dynamic modulus (E') and dynamic loss modulus (E'') of hydrogels are measured using a TMA. A sample holder of TMA and schematic TMA curves of hydrogel applied sinusoidal oscillation in water are shown in Figure 2-13. Gel sample is dipped in water using a sample holder shown in A in Figure 2-13. Frequency ranges from 0.01 to 20 Hz. Applied stress depends on rigidity of gel. Ordinarily, ca. 1×10^3 Pa is applied. Measurements are carried out for several minutes at each temperature. From

Lissajous diagram, E', E'' and tan δ are calculated using the following equations.

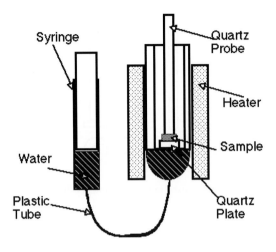

Figure 2-12. Schematic conformation of sample cell for the measurement of swelling of sample in water.

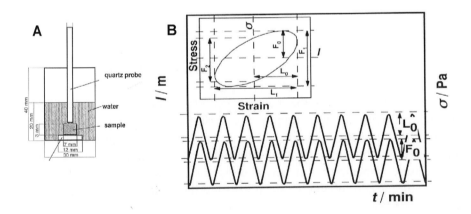

Figure 2-13. TMA sample holder measuring hydrogels in water (A) and schematic TMA curves of hydrogel applied sinusoidal oscillation in water (B). Upper left column shows Lissajous diagram.

$$|E*| = \frac{1}{A}\left(\frac{F_1}{L_1}\right)$$

(2.1)

$$E'' = \frac{1}{A}\left(\frac{F_2}{L_1}\right) \qquad (2.2)$$

$$sin\delta = \frac{F_1}{F_2} \qquad (2.3)$$

$$E' = |E^*|\cos\delta \qquad (2.4)$$

$$tan\delta = \frac{E''}{E'} \qquad (2.5)$$

Where $|E^*|$ is complex modulus and A is cross sectional area.

1.4 Dynamic mechanical analysis (DMA)

Viscoelastic properties of green polymers in solid state have been investigated by various techniques for about 50 years. Dynamic modulus (*E'*), dynamic loss modulus (*E''*) and tan δ are measured as functions of temperature and frequency by forced oscillation method. Torsion braid analysis is also used for samples which are difficult to make into films or fibres. Although various types of apparatuses are used, conformation of a representative apparatus for the measurement of viscoelasticity in green polymers in the solid state is shown in Figure 2-14. It is necessary to investigate green polymers having hydrophilic groups in humid conditions. In order to measure the viscoelastic properties in humid conditions, self-made and commercial apparatuses are used. By using the humidity controlling apparatuses, relaxations can be measured as functions of both relative humidity and temperature by computer control [48, 49]. A self made apparatus capable of measuring the sample in water is also reported [50]. Mathematical basis of viscoelasticity can be found elsewhere [51, 52].

The following information can be obtained by viscoelastic measurements of green polymers.

1. Dynamic modulus, dynamic loss modulus and tan δ as function of temperature and frequency
2. Temperature of the main chain relaxation (glass transition)
3. Temperature of local mode relaxations

4. Activation energy of each relaxation

An example of experimental conditions for standard measurements of green polymers by viscoelastic measurements is as follows; temperature range; 120 - 470 K, heating and cooling rate; 0.5 - 2 K min^{-1}, frequency; 0.1 - 200 Hz.

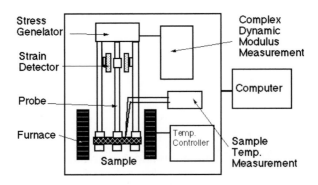

Figure 2-14. Example of conformation of apparatus for the measurement of viscoelasticity of green polymers in the solid state.

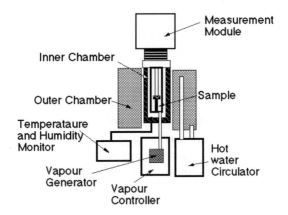

Figure 2-15. Conformation of apparatus for the measurement in humide conditions.

In order to measure the viscoelasticity of solid green polymers in humid conditions, various extra items of equipment have been made in the laboratory. Recently, apparatuses capable of changing relative humidity at a temperature from ca. 273 to 360 K are commercially available. Conformation of a humidity controllable apparatus is shown in Figure 2-15 [53].

2.   OTHER CHARACTERIZATION METHODS

In this section, other characterization methods which have been used to investigate the physicochemical properties of green polymers are briefly described.

2.1   Mechanical tests

2.1.1   Standard measurements

Stress-strain tests according to the industrial standards at room temperature in atmospheric conditions were carried out using an Instron type mechanical testing machine in order to measure the mechanical properties of green polymers. Force was applied by either elongation or compression methods. The tensile method was applied to films, sheets and plates and the compression method to foams and composites. When the sample amounts prepared in the laboratory were not sufficient to fit the conditions indicated by industrial standards, the conditions were modified according to the situation. Experimental conditions used in this book are as follows; size of test specimen are 20 to 100 mm for elongation tests and 20 x 20 x 20 mm for compression test. Cross head speed was 0.5 mm min^{-1}.

Figures 2-16 shows schematic stress-strain curve of compression test. In this book, strength of compression (σ_c MPa) was defined as the highest load (f_m) in the linear part of stress-strain curve for the compression test. Stress of yielding point (σ_y) is the maximum stress in stress-strain curve.

$$\sigma_c = \frac{f_m}{\sigma_0} \tag{2.6}$$

$$\sigma_y = \frac{f_y}{\sigma_0} \tag{2.7}$$

where σ_0 is cross section area of the specimen. Compression modulus is defined as the gradient of the linear part of stress-strain curve

$$E_c = \frac{\Delta\sigma}{\Delta\varepsilon} \tag{2.8}$$

where $\Delta\sigma$ and $\Delta\varepsilon$ are shown in Figure 2-16.

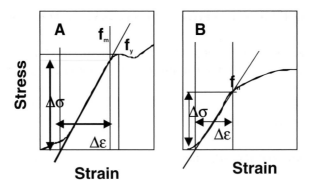

Figure 2-16. Schematic stress-strain curves.

2.1.2   Humidity controlled measurements

Due to hydrophilic characteristics of green polymers, the stress-strain measurements in an atmosphere whose humidity is controlled are important in obtaining reliable results. In order to carry out the above experiment without an attached sample probe, a simple method has been developed. (1) Samples in film, fibre or fabric, are put in a large size polyethylene envelop and maintained in a dessicator whose humidity is controlled using a saturated aqueous solution of inorganic salt (ref 2.2.2, Table 2-1). At this stage, the envelope containing samples is fully open. (2) After maintaining the sample for one day or more, samples and envelope are quickly taken out from the dessicator and the envelope is sealed with adhesive tape, (3) The sample is set in a tester together with the envelope. At this stage, the envelope is loosened in order to have sufficient room after stretching or compressing the sample. Using the above method, if room temperature is constant, the exact amount of water restrained by the sample can be calculated by referring to the data of sorption isotherm.

2.2   Measurements of basic physical properties

2.2.1   Density

Apparent density

Volume of the samples is measured using a caliper or other measuring equipments. Sample mass is obtained using a weighing balance. In order to obtain reliable data, the number of the specimens should be more than three and significant figures of size and those of mass should be coordinated.

Density

Density of the sample is measured using an Ostwald type picnometer or density column method depending on characteristics of the samples. In the above methods, organic solvents or aqueous solution of inorganic salts are necessarily used. Green polymers, either in original or extracted form, easily retain a certain amount of water. Air trapped in the micro-order structure of natural polymers is also difficult to exclude. Reorganization of the higher order structure when the sample is immersed in solvents must be taken into consideration. On this account, it is necessary to handle the sample carefully taking into consideration various factors in order to obtain reliable density data.

2.2.2   Water sorption

Although sorption experiments are a conventional method using only glass ware and a weighing balance, the obtained results still give us useful information, when the experiment is carefully carried out. For instance, (1) ca. 500 mg - 1 g of dry sample in a weighing bottle is placed in a dessicator maintained by an aqueous saturated solution of inorganic salt. Sample mass increases with increasing time and reaches a constant value. (2) The same experiment is carried out using different kinds of salts obtainable at various humidity levels. Typical inorganic salts which are ordinarily used in laboratories are listed in Table 2-1.

Table 2-1. Representative inorganic salts for preparing different relative humidites in a temperature range from 293 to 303 K

Materials	Relative humidity / %
$ZnCl_2$	9-11
CH_3COOK	20-23
$CaCl_2$	29-33
K_2CO_3	42-44
$Ca(NO_3)_2$	47-56
NH_4NO_3	59-67
NaCl	75-78
KCl	84-86
KNO_3	91-94

Data are quoted from Cellulose Handbook (Asakura Pub., Tokyo 2000)
Original data of the handbook were colleted from International Critical Table and reported papers published in the middle of the 20th century.

Sulfuric acid with various concentrations is also used for controlling the relative humidity. Predetermined relative humidity can freely be prepared by changing the concentration, when sulfuric acid is used. In ordinary experiments, inorganic salts are used, since green polymers easily degrade in sulfuric acid atmosphere. (3) The sample mass is weighed as a function of time at various humidity levels and saturated values are obtained. (4) From the relationship between relative humidity and mass of water at the equilibrium point, the number of molecules attaching to the first layer of the sample is calculated using the Brunauer-Emmett-Teller (BET) equation [2].

$$\frac{x}{v(1-x)} = \frac{1}{v_m c} + (c-1)\frac{x}{v_m c} \qquad (2.9)$$

where x is relative humidity, v total volume of gas, v_m volume of gas covering the monolayer area of matrix in monolayer and c is constant. The values x and v are experimentally obtained. Accordingly, when $x/[v(1-x)]$ is plotted against x, v_m and c can be calculated.

Ordinarily, the above experiments are carried out at room temperature, however temperature dependency is also measured. The most difficult point of sorption isotherm measurements is to prepare completely dry green polymers. In order to confirm whether a trace amount of water remains or not, it is useful to examine DTA or DSC heating curves. A large endothermic peak is observed when the sample contains a small amount of water, since the heat of vaporization is large.

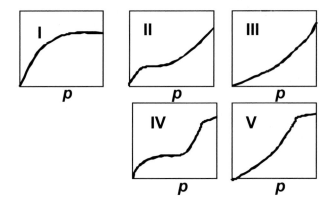

Figure 2-17. Water absorption pattern.

Water absorption pattern is ordinarily classified into several typical shapes as schematically shown in Figure 2-17. Five typical patterns (cases I, II, III, IV and V) are as follows, i.e. case I (Langmuir type absorption) is observed when water is sorbed in monolayer, case II (Sigmoid type absorption) is most frequently observed among organic materials and multilayer sorption is found at high RH, case III (Freundlich type absorption) is observed when water is sorbed on hydrophilic polymers and cases IV and V correspond to cases II and III showing capillary condensation at high RH range. The cases of IV and V are observed for porous materials, such as charcoal and silica gels. Due to the limitation of capillary volume, a large hysteresis is ordinarily observed. The above classification is not intrinsic for the materials, the absorption curve of hydrophilic polymers changes, when the crystallinity or higher order structure varies. Experimental technique of water absorption measurement is simple, but useful information can be obtained, although careful sample handling is necessary in order to obtain reliable data. A typical sorption isotherm of cellulose is shown in Figure 2-18. Molecular interpretations of sorption in green polymers are found elsewhere [55-57].

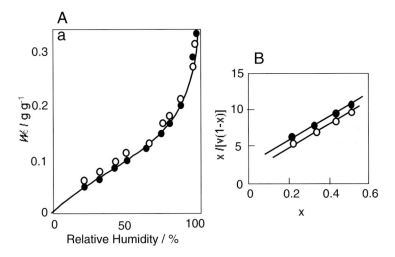

Figure 2-18. Typical sorption isotherm of cellulose at 298 K and BET plot. A: sorption isotherms. Open and filled circles show 1st and 2nd experiments, B: Brunauer-Emmett-Teller (BET) plot (Brunauer, S., Emmett, P. H. and Teller, E., 1938, *J. Am Chem. Soc.* **60**, 309).

2.2.3   Swelling ratio

Samples with known mass and dimension are immersed in water or organic solvents under pre-determined conditions. After the equilibrium swelling is attained, swelling ratio and water content are calculated.

$$\text{Swelling ratio} = \frac{V_{wet}}{V_{dry}} \times 100, \% \qquad (2.10)$$

Where V_{wet} is volume in the wet state and V_{dry} that in the dry state. Sample mass is frequently used for calculation of the swelling ratio, since the volume of the sample is difficult to measure. When the solvent in the gel is large, volume was calculated using known density of the solvent. Since evaporation of solvents occurs during mass measurements, the dimension of samples strongly depends on experimental conditions. On this account, the significant figures of swelling ratio are not large.

When the shape of gel sample is irregular, the tea bag method is used in order to calculate the swelling ratio at various temperatures. Polyester nonwoven fabric (thickness 0.2 mm) with size of 40 x 80 mm was sealed at 420 K. Weighed gel sample was inserted in the bag and immersed in water controlled at various temperatures. The swollen gel was weighed and the swelling ratio was calculated after the mass of tea bag was calibrated.

Swelling ratio in one direction of the sample can be measured using a microscope equipped with a micro scale. It should be noted that the linear swelling ratio depends on sample direction. As described in 1.3, TMA is used for the measurements of linear swelling ratio in solvents. The equilibrium swelling ratios obtained by visual observation accorded well with those obtained by TMA data, when applied stress is extrapolated to 0 at the equilibrium condition [48].

2.2.4   Viscometry

Viscometry has been extensively utilized in order to confirm the gel-sol transition of green polymer hydrogels. A cone plate type viscometer equipped with temperature control apparatus is ordinarily used. Experimental conditions which are usually used for the investigation of aqueous solution of polysaccharides are as follows, concentration = 1 - 2 %, parallel plate = 25 or 40 mm, frequency sweep = 0.1 - 200 rad sec^{-1}, and torque = 0.2 - 200 μ Nm, temperature = 278 - 333 K.

2.2.5   Falling ball method

In order to determine the gel-sol transition temperature, the following procedure has been used (1) the gel is prepared in a test tube, (2) a small steel ball with a diameter of several mm is inserted in the upper part of the gel, (3) the test tube is placed in a water bath whose temperature is isothermally controlled, (4) read the height of the steel ball using an appropriate device, such as a cathetometer, (5) temperature of the bath increases at 1 K min^{-1} or at a slower rate, (6) read the height of steel ball as a function of temperature, (7) the steel ball falls abruptly when the sol-gel transition occurs, (8) relationship between the height and temperature is obtained. (9) the gel-sol transition is determined at the cross point of the extrapolated line from the height of flat gel state and gradient of the height decrease after transition. Reliable data can be obtained, although this is not a modern method, when the gel-sol transition temperature is compared with that obtained by other methods, such as DSC.

2.3   Microscopy

2.3.1   Polarizing light microscopy

A polarizing light microscope equipped with a temperature controlling system is useful for the observation of green polymers which tend to form liquid crystals in dry or wet conditions. An objective lens with long distance focus is necessary in order to prevent the temperature increase of the lens from the heating cell.

2.3.2   Scanning electron microscopy

A scanning electron microscope equipped with dry or cryo-cell is used for the observation of green polymers and polymers derived from green biomass. The surface was spattered with gold. The standard technique was used [58].

2.3.3   Atomic force microscopy

The atomic force microscopic image shown in this book was obtained by tapping mode in the air at room temperature. Water soluble samples in a concentration ranged from 1×10^{-3} to 1×10^{-5} wt % were spread on a freshly cleaved mica surface.

2.4   Spectroscopy

2.4.1   Infrared spectroscopy

Both Fourier transform infrared (FTIR) and grating infrared (IR) spectrometers were used. A heating cell designed for obtaining the data of temperature dependency of absorption bands in this book is shown in Figure 2-19. A thin film sample was sandwiched between rock NaCl crystals and the temperature was increased slowly so as not to break the window of the NaCl crystals. Temperature was varied from ca. 300 to 480 K.

Absorption band is schematically shown in Figure 2-20. Baseline optical density (BOD) was calculated using equation 2-11. In order to compare the intensity among different samples, relative optical density was calculated using a characteristic absorption band as an internal standard.

$$\text{Baseline optical density} = \log\frac{AB}{CB} \qquad\qquad (2.11)$$

IR sample holder for H to D exchange of green polymer films is shown in Figure 2-21. Using a vacuum glass line, D_2O vapour was exchanged in the sample holder. The change of IR absorption, after D_2O vapour was introduced into the cell, was monitored as a function of time.

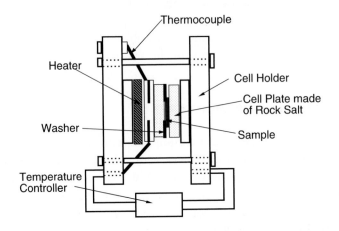

Figure 2-19. Sample holder for heating.

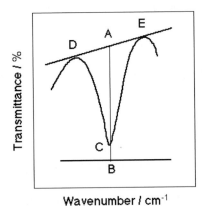

Figure 2-20. Schematic IR absorption band to calculate the baseline optical density.

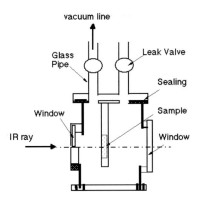

Figure 2-21. Sample holder for H to D exchange of polymer films.

2.4.2   Nuclear magnetic relaxation

Nuclear magnetic relaxation was measured using a Fourier transform nuclear magnetic resonance spectrometer. The basic concept and mathematical basis are found elsewhere [59, 60]. Longitudinal relaxation time (T_1) was measured by 180-τ-90 pulse method and transverse relaxation time (T_2) was measured by Meiboom-Gill variant Curr-Purcell method and free induction decay method depending on the relaxation time. Temperature was controlled from 120 K to 470 K. Temperature was changed stepwise.

2.4.3   Electron spin resonance spectrometry

In order to calculate quantitatively the number of spin, a double mode cavity was used when temperature dependency was measured. 1,1-Dipheyl-2-picrylhydrazyl (DPPH) was used as a standard and concentration was calibrated at each temperature.

2.5   X-ray diffractometry

2.5.1   Wide angle x-ray diffractometry (WAX)

A standard wide angel x-ray diffractometer equipped with a temperature controlling apparatus was used. In order to measure the green polymers in wet conditions, hand made sample holders were used, (1) a hole with diameter several mm was made in a metal plate with thickness ca. 1 mm, (2) both sides of the hole were windowed with thin mica films using an adhesive. When mica is not appropriate due to the short time interval between sample preparation and measurement, polyimide films were used as the material of the windows, and silicone grease was used as an adhesive. Either mica or polyimide film was measured under the same conditions of sample measurements, and the background scattering was extracted from the sample curve. Temperature changed stepwise or was heated at a constant rate.

The x-ray diffraction of green polymers during water sorption was measured using a sample holder equipped with moisture generator. Temperature of the sample holder was maintained at a constant and relative humidity was varied from 5 to 100 % RH. Schematic conformation of self made apparatus is shown in Figure 2-22.

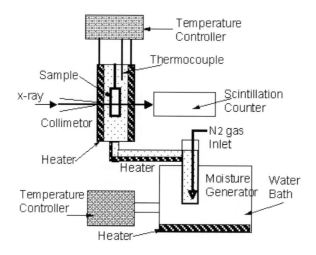

Figure 2-22. Sample holder for H to D exchange of polymer films.

2.5.2 Small angle X-ray diffractometry (SAX)

Small angle x-ray diffractometry of polysaccharide hydrogels was carried out using scattering orbital radiation, SOR (Tsukuba High Energy Physics Centre). Samples were placed in a hand made sample cell whose conformation is similar to that described in 2.5.1. Measurement conditions are described in the captions of figures when SAX was used.

REFERENCES

1. Slade, Jr P. E. and Jenkins L. T., Ed. Techniques and Methods of Polymer Evaluation, Marcel Deckker, New York, 1966, vol. 1 *Thermal Analysis*, 1970, vol 2 *Thermal Characterization Techniques*.

2. Porter, R. S. and Johnson J. F., Ed. 1974, *Analytical Calorimetry*. Prenumm, 1968, (vol. 1), (vol 2).
3. Mackenzie, M. C. Ed., 1972, *Differential Thermal Analysis*, London, Academic Press, 1970, vol 1, vol 2.
4. Wendlandt, W. W., 1975, *Thermal Method of Analysis*. 2nd Ed. New York, John, 1974.
5. Kambe, H. ed. *Thermal Analysis. Kodansya*, Tokyo (Japanese).
6. Keattch, C. J, Dollimore D., 1975, *An Introduction of Thermogravimetry*. 2nd Ed. London, Hyden.
7. Chen, J., 1981, Thermal Analysis and its Application to Ceramics. Beijing, Chinese Building Industry Press (Chinese).
8. Kambe, H. ed. Liu, Z. transl. 1982, *Thermal Analysis*. Beijing, Chemical Industry Press, (Chinese).
9. Mrevlishvili, G. M. 1984, *Low Temperature Calorimetry of Biopolymers*. Tobilishi, Metznieva, (Russian).
10. Zhang, Z. et al. 1984, *Thermal Apparatus in Analysis*, Beijing, Machanical Industry Press, (Chinese).
11. Soc. Thermal Analysis and Calorimetry Japan, ed. 1985, *Fundamental and Application of Thermal Analysis*. Tokyo, Kagaku Gijutsu Pub, (Japanese).
12. Chen, J., Li C., 1985, *Thermal Analysis and its Applications*. Beijing, Science Pub, (in Chinese).
13. Dodd, J. W. and Tonge K. H., 1987, *Thermal Methods*. Chichester, John Wiely.
14. Li, Y. 1987, *Thermal Analysis*. Beijing, Quing-Hua University Press, (Chinese).
15. Brown, M. 1988, *Introduction to Thermal Analysis*. New York, Chapman and Hall.
16. Soc. Thermal Analysis and Calorimetry Japan ed. 1989, *Fundamental and Application of Thermal Analysis* (2nd Ed.). Tokyo, Kagaku Gijutu Pub, (Japanese).
17. Wunderlich, B., 1990, *Thermal Analysis*. Boston, Academic Press.
18. Saito, A. 1990, *Fundamental of Thermal Analysis*. Tokyo, Kyoritsu, (Japanese).
19. Gao, J. et al. 1990, *Thermal Analysis Curves of Polymeric Materials*. Beijing, Science Press, (Chinese).
20. Xu, G. and Yuan J. et al. 1990, *Common Apparatus of Thermal Analysis*. Shanghai: Shanghai Science and Technology Press, (Chinese).
21. Charsley, E. L. and Warrington S. B. eds., 1991, *Thermal Analysis - Techniques and Applications.* Royal Soc Chem., Cambridge.
22. Liu, Z. ed., *General Thermal Analysis*. Chemical Industry Press, Beijing, 1991 (Chinese).
23. Cai, Z., *Thermal Analysis*. Higher Eduction Press, Beijing, 1991 (Chinese).
24. Kambe, H. and Ozawa, T. eds., 1992, *Thermal Analysis*. Tokyo Kodansya Scientific, (Japanese).
25. Mrevlishvili, G. M. Uedaira, H. and Ueadiar, Ha. transl. 1992, *Low Temperature Calorimetry of Biopolymers*. Hokkaido Univ Press, Sapporo, (Japanese).
26. Bershtein, V. A., Egorov, V. M. and Kemp T. J., transl. *Differential Scanning Calorimetry of Polymers*, Ellis Horwood, New York, 1994.
27. Soc Thermal Analysis and Calorimetry Japan ed. 1994, *Fundamental and Application of Thermal Analysis* (3rd Ed.), Realize Pub, Tokyo (Japanese).
28. Liu, Z. and Hatakeyama, T. eds. 1994, *Handbook of Thermal Analysis*, Beijing, Chemical Industry Press (Chinese).
29. Hatakeyama, T. and Quinn, F. X., 1994, *Thermal Analysis Fundamentals and Applications to Polymer Science*, John Wiley, Chichester.
30. Shen X., *Differential Thermal Analysis, Thermogravimetric Analysis and Kinetics of Isothermal Reaction in Solid State*. Metallurgical Industry Press, Beijing 1995 (Chinese).

31. Thermal Analysis Application Group ed., 1996, *Applied Thermal Analysis*. Tokyo, Daily Industry News Paper Pub. Tokyo (Japanese).
32. Turi, E. A. Ed. 1997, *Thermal Characterization of Polymeric Materials*. 2nd ed Academic Press, Orlando.
33. Hatakeyama, T. and Liu, Z., 1998, *Handbook of Thermal Analysis*. John Weily, Chichester.
34. Liu, Z. and Hatakeyama, T. and Zhang, X., 2001, *Thermal Measurements of Polymeric Materials*. Industrial Chemistry Press, Beijing, (Chinese).
35. Japan Society of Calorimetry and Thermal Analysis ed., 1998, *Calorimetry and Thermal Analysis, Maruzen* (Japanese).
36. Albac Rikoh. Ed. 2003, *Recent Advance in Thermal Measurements*, Agne Eng. Centre, Tokyo, (Japanese).
37. Arii, T., Senda, T., Kishi, A. and Fujii, N., 1995, *Thermochimica Acta*, **267**, 209.
38. Arii, T., Teramoto, K. and, Fujii, N., 1996, *J. Thermal Analysis*, **47**, 1649.
39. Momota, M., Ozawa, T. and Kanetsuna, H., 1990, 1998, Thermal analysis of polymer samples by a round robin method. Part V. Thermogravimetry, *Thermochimica Acta*, **159**, 125-137.
40. Hirose, S., Kobshigawa, K., Izuta, Y. and Hatakeyama, H., Thermal degradation of polyurethanes containing lignin studied by TG-FTIR. *Polym. Int.*, **47**, 347-356, ().
41. Nakamura, K., Nishimura, Y., Zetterlund, P., Hatakeyama, T. and Hatakeyema, H., 1996, TG-FTIR studies on biodegradable polyurethanes containing mono- and disaccharide components. *Thermochim. Acta*, **282/283**, 433-441.
42. Hatakeyama, H., Thermal analysis of environmentally compatible polymers containing plant components in the main chain. *J. Therm. Anal. Cal.*, **70**, 755-955 (2002).
43. Nakamura, S., Todoki, M., Nakamura, K. and Kanetsuna, H., 1988, Thermal analysis of polymer samples by a round robin method. I. Reproducibility of melting, crystallization and glass transition temperatures. *Thermochimica Acta*, **136** 163-178.
44. Hatakeyama, T. and Kanetsuna, H., 1989, Thermal analysis of polymer samples by round robin method Part II, Factors affecting heats of transition, *Thermochimica Acta*, **138**, 327-335.
45. Hatakeyama, T., Kanetsuna, H. and Ichihara, S., Thermal analysis of polymer samples by round robin method Part III, heat capacity measurement by DSC, *Thermochimica Acta*, **146**, 311-316 (1989).
46. Takahashi, T., Serizawa, M., Okino, T. and Kaneko, T., 1989, A round-robin test of the softening temperature of plastics by thermomechnical analysis, *Thermochimica Acta*, **147**, 387-399.
47. Nakamura, K., Kinoshita, E., Hatakeyama T. and Hatakeyama H., 2000, TMA measurement of swelling behavior of polysaccharide hydrogels, *Thermochimica Acta*, **352-353**, 171-176.
48. Yano S. and Hatakeyama H., 1988, Dynamic viscoelasticity and structural changes of regenerated cellulose during water sorption. *Polymer,* **29**, 566.
49. Yano S., 1993, Dynamic viscoelastic properties of carboxymethylcellulose during isothermal water sorption, *Polymer*, **34**, 2528-2532
50. Manabe, S., Iwata M., Kamide, K., 1986, Dynamic mechanical absorptions observed for regenerated cellulose solids in the temperature range from 280 to 600K. *Polym. J.* **18**, 1-14.
51. Gross, B., 1953, *Mathematical Structure of the Ttheories of Viscoelasticity*. Hermann & Cie, Éditeurs, Paris.

52. Yano, S. and Kitano, T., 1996, Dynamic Viscoelastic Porperties of Polymeric Matearials, in Handbook of Aapplied Polymer Processing Technology, Chapter 4, pp 125-188, N. P. Cheremisinoff and P. N. Gheremisinoff, eds. Marcel Dekker, Inc., New York.

53. Katoh, H. Nakamura, T. and Okubo, N., 1999, Dynamic mechanical measurements of polymer under the controlled moisture atmosphere, Netsu Sokutie **26**, 56-57 (Japanese).

54. Errede, L. A. 1991, *Molecular Interpretations of Sorption in Polymers*, Part I, Springer-Verlag, Berlin.

55. Hydration Processes in biological and macromolecular systems, Faraday Discussions, No. 103, 1996, The Royal Society of Chemistry, London.

56. Morra, M. Ed. 2001, *Water in Biomaterialas Surface Science*, Wiley, Chichester.

57. Goldstein J. I., Lyma E., Newbury D. E., Lifshin E., Echlin P., Sawyer L., Joy D. C. and Michael, J. R., 2003, *Scanning Electron Microscopy and X-ray Micronalaysis*, Kleuwer Academis/Plenum Pub. New York.

58. Farrar, T. C. and Becker, E. D., 1971, *Pulse and Fourier Transform NMR, Introduction to Theory and Methods*, Academic Press, New York.

59. Abragahm, A., 1961, *The Principels of Nuclear Magnetsm*, Clarendon Press, Oxford.

Chapter 3

THERMAL PROPERTIES OF CELLULOSE AND ITS DERIVATIVES

1.   INTRODUCTION

Cellulose is the most abundant organic compound and a representative renewable resource. According to the statistical calculation of the Food and Agriculture Association, US, $3{,}270 \times 10^9$ m^3 of cellulose exists on the earth and 1 % of it is currently utilized. Cellulose can be obtained from various plants, such as trees, cereals, cotton, jute, ramie, hemp, kenaf, agave, etc. It is also known that some bacteria produce cellulose. Cellulose separated from the above plants has been used as paper, textile, foods and fine chemicals. The chemical structure of cellulose is poly (β-1,4 D glucose) as shown in Figure 3-1 [1-3].

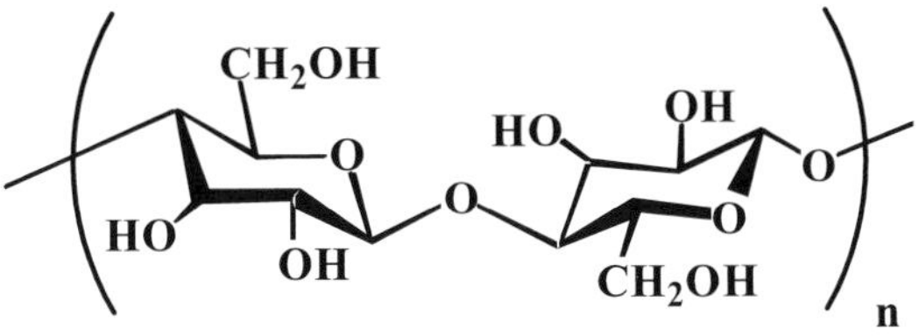

Figure 3-1. Chemical structure of cellulose.

Molecular size and its structural hierarchy is shown in Table 3-1. The molecular sizes shown in the table are not exact values, since molecular mass depends on the extraction method from living organs. Cellulose obtained directly from plants is categorized as natural cellulose, and once solved in various kinds of solvent is known as regenerated cellulose. Polymorphic structures are found in cellulose and cellulose derivatives. The crystalline structure of natural cellulose is roughly categorized as cellulose I, and that of regenerated cellulose as cellulose II. Recent studies on crystallography of cellulose suggest that cellulose I consists of two kinds of crystal, Iα and Iβ. The complex crystalline structure of natural cellulose is shown in Figure 3-2.

Table 3-1. Size of cellulose in each hierarchy

Hierarchy	Size
Molecule	0.33×0.39 nm^2
Micelle	5.0×6.0 nm^2
Micro-fibrill	25.0×25.0 nm^2
Fibrill	0.4×0.4 mm^2
Lamellae	~12.6 mm^2
Cell (Cotton)	~ 314 mm^2

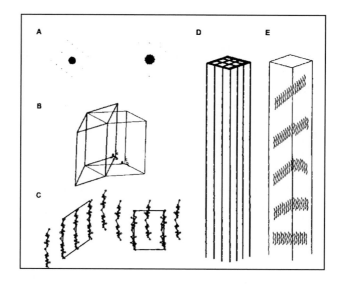

Figure 3-2. Crystalline structure of cellulose I_α and I_β [3].

Polymorphism of cellulose crystals and its mutual transformation are briefly summarized in Figure 3-3. In this figure, the left column shows the cellulose-I family and the right cellulose-II family. The crystalline structure of cellulose has been investigated for the past 80 years, however, discussion still continues among scientists. Concerning the details of the historical background, representative papers are cited in the references [4-18]. Crystallinity of natural cellulose depends on the original plants. The values of crystallinity also vary according to measurement methods, such as x-ray diffraction analysis, infrared spectroscopy and thermal analysis. Completely amorphous cellulose can be prepared by saponification of cellulose triacetate or mechanical grinding. Amorphous cellulose is used as a reference material.

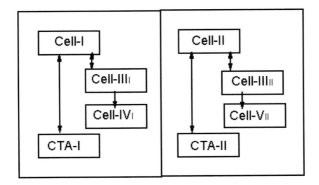

Figure 3-3. Polymorphism of cellulose crystal, Cell: cellulose, CTA: cellulose triacetate.

Cellulose derivatives have been synthesized for the past 100 years based on industrial demands [19]. Cellulose esters and ethers are the major derivatives. Representative derivatives, whose thermal analysis has been carried out, are shown in Tables 3-2 and 3-3 together with their chemical structures. In this chapter, thermal properties of natural and regenerated cellulose and derivatives are described.

Table 3-2. Representative cellulose derivatives (cellulose esters)

Cellulose Ester	Chemical Structure
Cellulose nitrate	$Cell\text{-}ONO_2$
Cellulose phosphate	$Cell\text{-}OPO_2Na_2$
Cellulose xanthate	$Cell\text{-}OCS_2Na$
Cellulose sulfate	$Cell\text{-}OSO_3Na$
Cellulose acetate	$Cell\text{-}OCOCH_3$

Table 3-3. Representative cellulose derivatives (Cellulose ether)

Cellulose ether	Chemical sturcture
Carboxymethylcellulose	Cell-OCH$_2$COONa
Methylcellulose	Cell-OCH$_3$
Ethylcellulose	Cell-OCH$_2$CH$_3$
Hydroxypropylcellulose	Cell-OCH$_2$CH(OH)CH$_3$

2.   THERMAL PROPERTIES OF CELLULOSE IN DRY STATE

2.1   Heat capacities of cellulose

When dry cellulose is heated from 120 to 470 K by DSC, no first-order phase transition is observed [20]. On this account, in DSC curves, only flat sample baselines can be obtained. The free molecular motion of the main chain of cellulose is restricted due to inter-molecular hydrogen bonding. Cellulose is insoluble in water, however, it sorbs a characteristic amount of water. Since the hydroxyl groups form hydrogen bonding with water molecules, it is difficult to obtain completely dry samples. If cellulose sorbing a slight amount of water is measured by DSC, a large endothermic peak attributable to vaporization of water is observed in a temperature range from 273 to 400 K. Peak temperature of vaporization depends on the amount of water. Since heat of vaporization is large (1339 J g^{-1} at 293 K), the endothermic peak of vaporization is used as an appropriate index for detecting the residual water in cellulose after drying. Not only cellulose, but also natural polysaccharides show no first order phase transition, if they are in the dry state.

Although no phase transition is measured, heat capacity (C_p) can be calculated by DSC using a reference material whose C_p values have been determined by adiabatic calorimetry. Figure 3-4 shows C_p values of various kinds of cellulose having different crystallinity. Amorphous cellulose shown in this figure was prepared by saponification of cellulose triacetate.

Cellulose triacetate film was immersed in NaOH dehydrated ethyl alcohol. By substitution of the acetyl group to the hydroxyl group in dehydrated condition, the structure of cellulose molecules is solidified in random arrangement maintaining the intermolecular space occupied by bulky acetyl side chains. Saponified samples show a typical halo pattern having an amorphous structure when measured by x-ray diffractometry. Other cellulose samples shown in Figure 3-4 were in powder form. Crystallinity was calculated using an x-ray diffractogram in a 2θ range from 5 to 40 degrees (Table 3-4).

Table 3-4. Crystallinity of various kinds of cellulose

Natural cellulose	Crystallinity (%)	Regenerated cellulose	Crystallinity (%)
Hemp yarn	69	Polynosic rayon	46
Cotton yarn	54	Cupra rayon	43
Cotton lint	52	Viscose rayon	42
Wood cellulose	44		
Jute	36		
Kapok	33		

*Crystallinity was calculated using amorphous cellulose as a reference material.

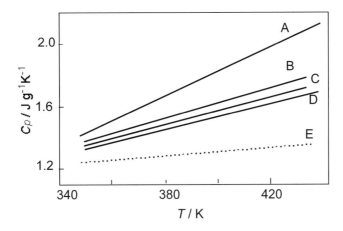

Figure 3-4. Heat capacities of various kinds of cellulose. A: amorphous cellulose, B: wood cellulose, C: jute, D: cotton, E: calculated data of cellulose with 100 % crystallinity. Power compensation type DSC (Perkin Elmer). Reference material; sapphire, Sample pan, open type aluminium. Sample mass = ca. 7 mg, heating rate = 10 K min^{-1}, N$_2$ flowing rate = 30 ml min^{-1}, Sample shape; powder was compressed in a pellet shape in order to come into contact tightly with the surface of the sample pan. Amorphous cellulose film was prepared by saponification of cellulose triacetate. The film was annealed at 460 K for 5 min [21].

As shown in Figure 3-4, C_p values increase linearly with increasing temperature. At the same time, C_p values decrease with increasing crystallinity of cellulose. If crystallinity is known, C_p values at an appropriate temperature can be calculated using simple additivity law.

$$Cp = XcCpc + (1 - Xc)Cpa \qquad (3.1)$$

where X_c is crystallinity, C_{pc} is C_p value of completely crystalline cellulose and C_{pa} is that of amorphous cellulose. C_{px} can be obtained by extrapolation as shown in Figure 3-5.

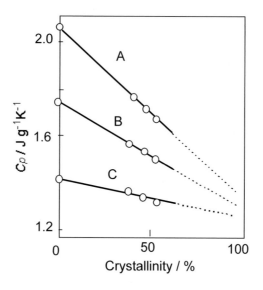

Figure 3-5. Relationship between crystallinity and heat capacities of cellulose at various temperatures. A: 430 K, B: 390 K, C: 350 K [21].

2.2   Glass transition of cellulose acetates with various degrees of substitution and molecular mass

Among various types of cellulose derivatives shown in Tables 3-2 and 3-3, cellulose acetate is widely used for practical purposes, such as photographic film, packaging materials, separating membranes etc. Cellulose acetate (CA) is ordinarily prepared from wood pulp by acetylation in acetic acid and sulfric acid. Chemical structure of CA is shown in Figure 3-6. In this figure, R is the acetyl group. Degree of substitution (DS) is defined as the number of the acetyl groups substituted from the hydroxyl group. As an industrial index, CA samples with DS ranged from 2.4 to 2.56 are designated as cellulose diacetate (DCA) and those from 2.8 to 2.92 as cellulose triacetate (CTA). It is known that the C6 position is preferentially substituted, and the substitution of 2C and 3C occurs statistically. The position of the acetyl group can be determined by nuclear magnetic resonance spectroscopy (NMR).

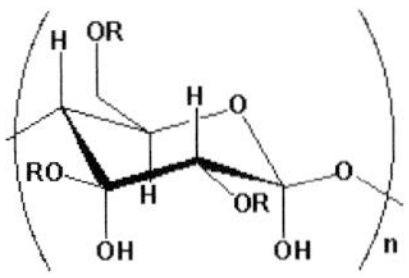

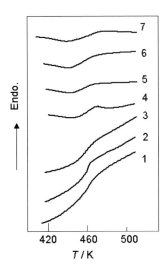

Figure 3-6. Chemical structure of cellulose acetate. R: $COCH_3$ or H.

Since cellulose acetates are soluble in organic solvents, such as chloroform, it is possible to prepare fractionated samples with different molecular mass by successive precipitation. Figure 3-7 shows representative DSC heating curves of cellulose acetate fractions with different molecular mass. When molecular mass increases, thermal decomposition starts immediately after completion of melting or glass transition [22].

Figure 3-7. Representative DSC heating curves of fractions of cellulose acetate with degree of substitution 2.92. M_v 1: 4.7×10^4, 2: 1.97×10^5, 3: 2.22×10^5, 4: 3.59×10^5, 5: 4.56×10^5, 6: 5.83×10^5, 7: weight-average molecular weight 2.35×10^5, Experimental conditions; the viscosity-average molecular weight was estimated using the Mark-Houwick-Sakurada equation at 298 K. N,N-dimethylacetamide was used as a solvent. Power compensate DSC (Perkin Elmer), N_2 flow rate = 30 ml min^{-1}, heating rate = 10 K min^{-1}.

Relationship between T_g and M_v is shown in Figure 3-8. T_g increases with increasing molecular weight. When the degree of substitution decreases, T_g maintains a constant value regardless of molecular weight [23].

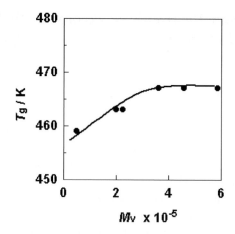

Figure 3-8. Relationship between T_g and M_v of cellulose acetate with degree of substitution 2.92. M_v: viscosity average molecular mass. Experimental conditions; see Figure 3-7 caption.

Figure 3-9 shows the relationships between T_g estimated by DSC heating curves of CA with various DS's and molecular weight. As shown in this figure, when the degree of substitution decreases, glass transition temperature (T_g) maintains a constant value regardless of molecular weight and only depends on degree of substitution. With increasing degree of substitution, T_g decreases due to expansion of intermolecular distance.

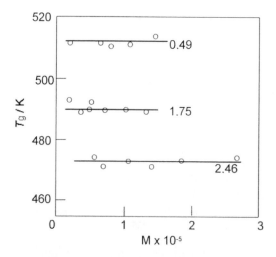

Figure 3-9. Relationship between T_g and M_v of cellulose acetate with different degree of substitution. Numerals in the figure show degree of substitution.

2.3　　Heat capacity of sodium carboxymethylcellulose with different molecular mass and degrees of substitution

Sodium carboxymethylcellulose (NaCMC) is a representative water soluble polyelectrolyte derived from cellulose (see Table 3-3). Figure 3-10 shows the chemical structure of NaCMC. When the carboxymethyl groups are introduced into cellulose, the higher order structure of cellulose gradually changes [23]. As shown in Figure 3-11, the crystallinity of carboxymethy-lcellulose (CMC) in acid form decreases with increasing number of carboxymethyl group, since inter-molecular distance increases due to bulky side chains. CMC's substituted by a monovalent cation salt are water soluble, however when divalent cations are substituted, water insoluble gels are formed. Among various kinds of CMC derivatives, sodium CMC is most widely utilized in various fields, as a glue for dying and weaving in the textile industry, a viscosity controlling compound in the food industry and an anti-deposition agent for detergent in the cleaning and cosmetic industries.

Figure 3-10. Chemical structure of carboxymethylcellulose (CMC). R= H or CH_2COOH.

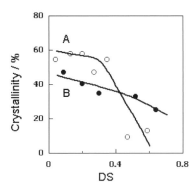

Figure 3-11. Relationship between crystallinity and degree of substitution of carboxymethyl-cellulose (CMC) in acid form. DS: total degree of substitution, A: natural cellulose (cotton), B: cellulose II (cupra rayon).

Figure 3-12 shows C_p curves of NaCMC with various molecular weights. Degree of substitution is 1.4. As shown in Figure 3-13, T_g values maintain a constant, while in contrast ΔC_p values decrease with increasing M_v, suggesting that molecular enhancement of NaCMC is depressed when molecular weight increases.

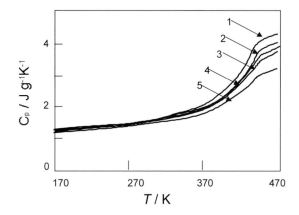

Figure 3-12. Heat capacity curves of sodium carboxymethylcellulose (degree of substitution = 1.4) with various molecular weights (M_w). 1: 1.7 x 10^4, 2: 3.4 x 10^4, 3: 5.9 x 10^4, 4: 1.03 x 10^5, 5: 3.8x 10^5 (See Table 3-5) Experimental conditions; Heat-flux type DSC (Seiko Instruments DSC220), heating rate 10 K min^{-1}, Reference material; sapphire samples were heated up to 373 K and maintained for 10 min in order to eliminate residual water in the sample, cooled to 170 K and heated [24].

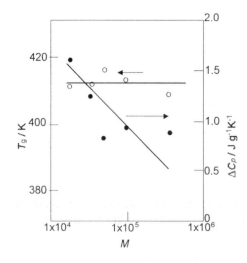

Figure 3-13. Relationship between glass transition temperature (T_g), heat capacity gap at T_g (ΔC_p) and molecular mass of NaCMC (DS = 1.4) [24]. Definition of ΔC_p (see Figure 2.10).

Figure 3-14 shows heat capacity curves of sodium carboxymethyl-cellulose with various degrees of substitution. When inter-molecular distance increases by the introduction of carboxymethyl groups, C_p values increase. It is also seen that ΔC_p values increase with increasing DS. When DS ranges from 0.6 to 0.8, ΔC_p values were ca. 0.75 J g^{-1} K^{-1} and when DS ranges from 1.4 to 1.7, ΔC_p values range from ca.1.25 to 1.30 J g^{-1} K^{-1}, respectively.

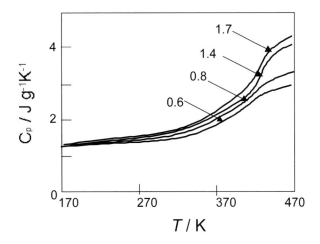

Figure 3-14. Heat capacity curves of sodium carboxymethylcellulose with various degrees of substitution. Numerals shown in the figure are DS (M_w=3.4 x 10^4).

Table 3-5. Molecular mass and degree of substitution of sodium carboxymethylcellulose

Degree of substitution (DS)	M_w
0.6	3.9 x 10^4
0.8	4.9 x 10^4
1.4	1.7 x 10^4, 3.4 x 10^4, 5.9 x 10^4, 1.03x 10^5, 3.8x 10^5
1.7	5.7 x 10^4

2.4 Hydrogen bonding formation of amorphous cellulose in dry state

As described in 3.1, natural cellulose is a crystalline polymer whose crystallinity ranges from ca. 20 to 90 %. The crystallinity of natural cellulose depends on plant species, for example the crystallinity of jute is ca. 70 %, whereas that of kapok is ca. 30 %. Crystallinity is ordinarily determined by x-ray diffractometry, infrared spectrometry and solid state NMR. In the initial stage of the investigation of x-ray diffractometry of cellulose, it was necessary to prepare amorphous cellulose as a reference in order to calculate crystallinity. Amorphous cellulose has mainly been prepared by two methods, i.e. one is milling using a ball mill by which fine powder can be obtained. The other is saponification of cellulose triacetate in dehydrated conditions. By saponification, the bulky side chains are converted into hydroxyl groups and the space of side chains is fixed, if no water molecules exist is the reaction system. In the experimental procedure, metal sodium was solved in ethyl alcohol and sodium alcholate solution was used for the purpose. CTA films were immersed in the above solution for several hours, washed by dehydrated alcohol several times and kept in dehydrated conditions. Figure 3-15 shows wide line x-ray diffractograms of amorphous cellulose prepared by saponification of CTA. A broad peak is observed at 2θ=20 degrees [25-29].

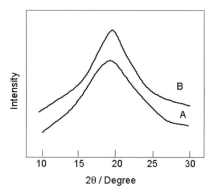

Figure 3-15. Wide line x-ray diffractogram of amorphous cellulose. A: original sample, B: pre-annealed sample.

When an amorphous sample obtained by the above procedure is heated by DSC in water eliminated conditions, a broad exotherm due to recombination of hydrogen bonding can be observed in a temperature range from 370 to 450 K [25-30]. The shape and enthalpy of exothermic peak vary when the structure of cellulose triacetate has been modified. When cellulose molecular chains are arranged in one direction, enthalpy of transition decreased since the inter-molecular bondings are easily formed. Figure 3-16 shows DSC heating curves of amorphous cellulose samples having various histories. As shown in the heating curves, once the sample is heated to 460 K and molecular rearrangement is completed, no transition can be observed although the x-ray diffractogram scarcely changed. At the same time, enthalpy of transition decreased when cellulose triacetate had been drawn before saponification.

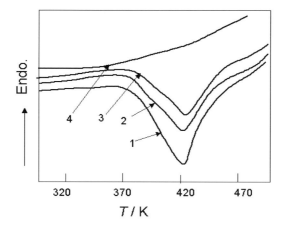

Figure 3-16. DSC curves of un-drawn (original) and drawn amorphous cellulose showing the effects of pre-drawing of cellulose triacetate before saponification, and annealed amorphous cellulose, 1: original (undrawn amorphous cellulose, 2: uni-axially drawn amorphous cellulose (draw ratio = 2 x 1), 3: bi- axially drawn amorphous cellulose (draw ratio = 2 x 5), 4: undrawn amorphous cellulose was annealed at 463 K. Samples; Cellulose triacetate was immersed in 1 % potassium hydroxide solution of dehydrated ethanol at room temperature for 24 hrs. Dehydrated ethanol was prepared by the use of calcium oxide and anhydrous calcium sulfate. The obtained samples were washed with dehydrated ethanol until the washing solution became neutral. Drawn amorphous cellulose was made using pre-drawn triacetate films. Two direction drawing was carried out, i.e. the second drawing was carried out perpendicular to the first one. Draw ratio was shown as a x b, where a is draw ratio of the first drawing and b is the second one. Measurement; Power compensation type DSC (Perkin Elmer), heating rate = 16 K min^{-1}, N$_2$ atmosphere, sample mass = ca. 8 mg.

When the original sample was annealed at a temperature where the exotherm was observed, the exothermic peak decreased depending on annealing temperature and time. Amorphous cellulose samples were maintained isothermally at a temperature range from 390 to 430 K for 60 min. At temperatures higher than ca. 430 K, the transition is completed too rapidly to monitor isothermal state. In contrast, at temperatures lower than 390 K, exothermic deviation on DSC curve is small enough to detect over a certain time. Figure 3-17 shows the enthalpy of transition at various temperatures. The leveling-off point indicated the apparent end of the exothermic process which was detectable by this method. The time for attaining the maximum enthalpy was found to decrease with increasing temperature. Isothermal change of IR spectra was also carried out and specific absorption band was correlated with DSC data, although the results are not shown here.

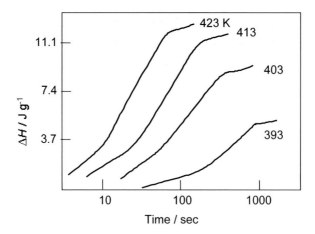

Figure 3-17. Isothermal changes of transition enthalpy of amorphous cellulose. Sample preparation; see Figure 3-16 caption. Measurement; Power compensation type DSC (Perkin Elmer). Temperature increased abruptly from room temperature to each pre-determined temperature and exothermic trace was detected as a function of time. The point where the baseline stabilized is defined as time = 0. After a certain period (ca. 60 min), total enthalpy was calculated and used for normalization.

Figure 3-18 shows the change of ratio of reacted (x) and non-reacted amount (a) with time at various temperatures. Apparently, observable molecular rearrangements appear to occur by a first-order mechanism. There is sufficient thermal motion of the cellulose chains to cause some concurrent alignment of a small segment of cellulose at primary nuclear site, which formed by hydrogen bonding in the initial stage. The differential equation defining this initiation process is as follows

$$ -\frac{dx}{dt} = k(a - x) \tag{3.2} $$

where k is rate constant independent of α, x is the amount of nuclei formed, and a is the amount of non-bonded part available for nucleation. The calculated rate constants are shown in Table 3-6. The apparent activation energy (E_a) for the primary nucleation process was calculated by using the Arrehenius relationship. In this case, E_a is assumed to be independent of temperature over the range cited. Activation energy was approximately 190 kJ mol^{-1}.

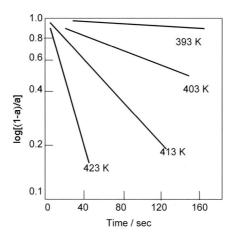

Figure 3-18. Relationships between log [(a-*x*)/a] and time (sec) of amorphous cellulose. x is the amount of nuclei formed, and a is the amount of non-bonded part available for nucleation.

Table 3-6. Calculated rate constant as a function of temperature

Temperature / K	Rate constant / sec^{-1}
393	6.4×10^{-4}
403	4.6×10^{-3}
413	1.3×10^{-3}
423	4.3×10^{-2}

The exothermic transition observed in amorphous cellulose is considered to consist of two processes, (1) the formation of hydrogen bond by free hydroxyl groups formed during saponification and (2) the formation of the crystallites composing nuclei for crystal growth, which could not be detected clearly by x-ray diffraction.

When amorphous cellulose is maintained in humid conditions, crystallization gradually starts and cellulose II type crystal is obtained. Crystallization of amorphous cellulose in humid conditions is described in section 2.7 of this chapter.

2.5 Glass transition of mono- and oligosaccharides related to cellulose

Phase transition behaviour of several representative mono- and oligosaccharides was investigated by DSC [30-32]. Figure 3-19 shows DSC heating curves of α-D-glucose monohydrate, α-D-glucose anhydride, β-D-glucose and cellobiose. When the samples were quenched in completely dry

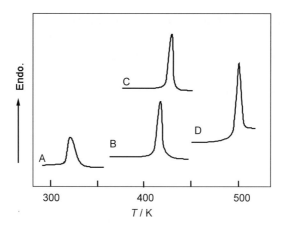

Figure 3-19. DSC curves of α-D-glucose monohydrate, α-D-glucose anhydride, β-D-glucose and cellobiose. A: α-D-glucose monohydrate, B: α-D-glucose anhydride, C: β-D-glucose, D: cellobiose. Measurements; Power compensation type DSC (Perkin Elmer), sample mass = 2 - 3 mg, heating rate = 1 K min^{-1}.

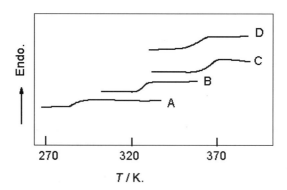

Figure 3-20. DSC heating curves of amorphous mono- and oligosaccharides relating to cellulose. A: D-glucose, B: cellobiose, C: cellotriose, D: cellotetriose. Sample preparation; The acetates of cellulose oligosaccharides were fractionated by ethanol-water gradient elution method. Charcoal-celite pretreated with 2.5 % stearic acid was used as a filler of the column. Each fraction of oligosaccharides was hydrolyzed after purification by rechromatography [33]. Measurements; Power compensation type DSC (Perkin Elmer), sample mass = 2 - 3 mg, heating rate = 10 K min^{-1}.

conditions, amorphous glucose and cellobiose were obtained. As shown in curve D in Figure 3-19, a baseline gap is observed before and after melting of cellobiose. This fact indicates that melting is masked by partial decomposition. Recrystallization is capable of taking place only when a trace amount of water is present.

DSC heating curves showing glass transition are also shown in Figure 3-20. When cellobiose was heated at a temperature higher than the melting peak, decomposition starts and the sample colour changes to light brown. On this account, cellobiose was quenched immediately after completion of melting in the DSC sample holder.

Figure 3-20 shows DSC heating curves of quenched glucose and oligosaccharides relating to cellulose. A baseline shift due to glass transition is observed. With increasing molecular weight, glass transition becomes difficult to measure.

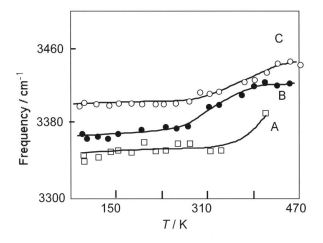

Figure 3-21. Relationship between frequency of OH stretching band at around 3400 cm^{-1} and temperature. A: D-glucose, B: cellotriose, C: cellopentaose, Measurements; see details, 2.4.1 in Chapter 2.

Figure 3-21 shows temperature dependency of OH stretching bands measured by infrared spectrometry. At around T_g measured by DSC, the shift of absorption bands can be observed [34].

Figure 3-22 shows heat capacities of amorphous D-glucose quenched from the molten state to glassy state, D-glucose anhydride and cellobiose. From this figure, it is clear that heat capacities of amorphous D-glucose are markedly high, suggesting the random molecular arrangement. Once amorphous glucose is formed in completely dry conditions, crystallization does not occur by annealing. If a trace amount of water is added to the amorphous cellulose, crystallization takes place.

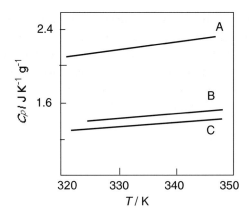

Figure 3-22. Heat capacities of amorphous D-glucose, D-glucose anhydride and cellobiose.
A: amorphous D-glucose, B: D-glucose anhydride. C: cellobiose.

3.   CELLULOSE-WATER INTERACTION

Phase transition behaviour of hydrated polymers has been widely investigated by various analytical techniques owing to the effect of water on the performance of commercial polymers and the crucial role played by water-polymer interactions in biological processes.   Mechanical and chemical properties of polymer change in the presence of a characteristic amount of water. At the same time, the behaviour of water is transformed in the presence of a polymer depending on the chemical and higher-order structure [35-39].

Water whose melting/crystallization temperature and enthalpy of melting/crystallization is not significantly different from that of normal (bulk) water is called freezing water. Those water species exhibiting large differences in transition enthalpies and temperatures, or those for which no phase transition can be observed calorimetrically, are referred to as bound water. Water fraction closely associated with the polymer matrix ordinarily shows no phase transition. Such fraction is called non-freezing water.   Less closely associated water fraction exhibit the melting / crystallization peaks and is referred to as freezing-bound water. The sum of the freezing-bound and non-freezing water fractions is the bound water content [38].

Although various methods, such as nuclear magnetic resonance spectroscopy (NMR), viscoelastic measurements and dielectric measurements are used in order to quantify the amount of bound water in hydrophilic polymers, TA is a technique characterized by various advantageous points, such as a small amount of sample mass, and a wide range of information on phase transition behaviour of water [38, 40-50].

In this section, cellulose- and cellulose derivative-water interaction investigated by TA are introduced. Phase transition behaviour of water attaching to the hydroxyl groups of cellulose is focused on. Water content and water concentration of the sample have been defined in various equations. In this book, water content (W_c) is defined as follows.

$$Water\ Content(W_c) = \left(\frac{m_{water}}{m_{sample}} \right) \tag{3.3}$$

where m_{water} is mass of water and is m_{sample} mass of dry sample

3.1 Phase transition behaviour of water restrained by cellulose

Since cellulose is the most important hydrophilic polymer, many authors have reported cellulose-water interaction using various experimental techniques and found that water restrained by cellulose has markedly different properties from free water [38]. It is known that melting and crystallization temperatures of the bound water in cellulose and other biopolymers are lower than those of free water. It has been considered that the molecular mobility of water is restricted on the polymer surface through the interaction with hydrophilic groups, and diffusion and penetration of water are retarded by the polymer matrix.

Figure 3-23 shows the DSC curves of bulk water (curve A) and water restrained by cellulose (curve B). When bulk water was cooled from 320 K to 150 K, the crystallization peak starts at around 260 K and in the heating curve, melting peak starts at 273 K. Temperature difference between ($T_{pm} - T_{pc}$) owing to super-cooling depends on scanning rate. Due to the above fact, melting enthalpy calculated from the heating curve is always larger than crystallization enthalpy. On this account, enthalpy obtained by cooling curve was calibrated taking into account the above difference. As shown in the cooling curve of curve II, a new small exotherm is observed at around 220 to 230 K (Peak II) together with the crystallization peak of water (Peak I). This peak is attributed to freezing bound water, which will be discussed in the latter section in detail. Melting peak starts at a temperature lower than 273K and shoulder peak can be seen in the low temperature side.

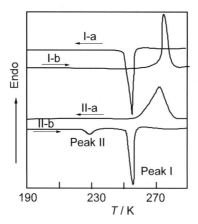

Figure 3-23. Schematic DSC heating and cooling curves of water restrained by celluloser. I: pure water, II: water restrained by cellulose $W_c = 0.8$ g g^{-1}, a: heating curve, b: cooling curve, Heat flux type DSC (Seiko Instruments), Scanning rate = 10 K min^{-1}.

Figures 3-24 (A) and 3-24 (B) show the stacked DSC cooling curves of water restrained by natural cellulose (cellulose I) and regenerated cellulose (cellulose II), respectively [24]. The broken line in Figure 3-24 (B) corresponds to the crystallization curve of bulk water. As shown in Figures 3-24 (A) and 3-24 (B, no crystallization peak was observed when W_c of cellulose I is below 0.15 g g^{-1} and that of cellulose II below 0.25 g g^{-1}. These facts suggest that below the above mentioned W_c's only non-freezing water exists in the cellulose I- and cellulose II-water systems. When W_c exceeded critical amounts for celluloses I and II, peak II was firstly observed.

The crystallization behaviour of water in cellulose II is complicated. When W_c is between 0.3 and 0.6 g g^{-1}, an intermediate peak (peak II') appears at a temperature higher than Peak II but lower than peak I as shown in Figure 3-24 (B). Peak II' shifts to the higher temperature side with increasing W_c, becoming a shoulder of peak I when W_c is over 0.5. Accordingly, it is considered that Peak II' also corresponds to the bound water [47]. When W_c exceeds a certain amount (0.19 g g^{-1} for cellulose I and 0.42 g g^{-1} for cellulose II), Peak I appears. However, when W_c is low, peak I appears at a temperature lower than that of the normal crystallization temperature of free water which is higher than that of bulk water. This suggests that Peak I is also under the influence of cellulose matrix.

Figures 3-25 (A) and 3-25(B) show the relationship between the peak temperature of crystallization peak of celluloses I and II. In the case of cellulose I, the temperature of Peak II is observed at 228 to 230 K at the low W_c region regardless of W_c. Peak I increases at the low W_c region. As shown

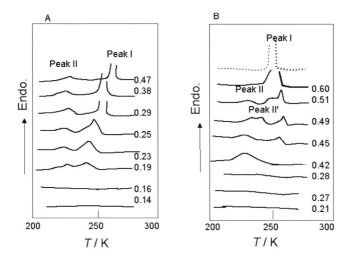

Figure 3-24. DSC heating curves of water restrained by natural and regenerated cellulose. Samples (a) natural cellulose (cotton linter, crystallinity estimated by x-ray diffractometry (x_c) = 52 %), (b) regenerated cellulose (rayon fibre, Asahi Chemical Co., x_c = 38 %), Numerals in figures show water content in g g^{-1}. Experimental conditions; Power compensation DSC (Perkin Elmer), sample mass = 3 - 5 mg, scanning rate = 10 K min^{-1}, sample pan; aluminium sealed type, temperature was calibrated using pure water. Starting temperature of melting of pure water was defined as 273 K. Preparation of water containing sample; The dry cellulose sample was weighed quickly and water was added using a microsyringe. After evaporation of excess water, the sample pan was sealed and weighed. W_c was calculated using Eq. 3.2. After sealing, the sample pan was placed in a heat oven for 1 h at 333 K, then maintained at 295 K for 24 hrs and weighed again [47].

in Figure 3-24 (B), peaks II and II' of cellulose II vary in a complex manner. Peak II shows a maximum at W_c = ca. 0.45 g g^{-1} and the intermediate peak separated from peak II at W_c = ca. 0.3 g g^{-1} and merged into Peak I at around W_c = 0.6 g g^{-1}. The above results suggest that the structure of amorphous region of celluloses I and II is quite different and molecular conformation successively changes with increasing water content.

Melting endotherms of water restrained by various W_c's are ordinarily a broad peak with shoulder peak in the low temperature side or with no clearly detectable side peak. The starting temperature shifts to the high temperature side with increasing W_c. Figure 3-26 shows the peak temperature of the main melting peak as a function of W_c. The peak temperature increases linearly with increasing W_c up to W_c = 0.50 g g^{-1} and then levels off at 275 K, which agrees well with the peak temperature of bulk water as indicated by the chain line. As shown in Figure 3-26, DSC curves were not shown, and the melting peak becomes sharper with increasing W_c, i.e. the temperature difference between melting peak (T_{pm}) and starting temperature of melting

(T_{mi}', ref Figure 2-7 in Chapter 2) gradually decreases with increasing W_c. The temperature difference is far larger than the case of bulk water.

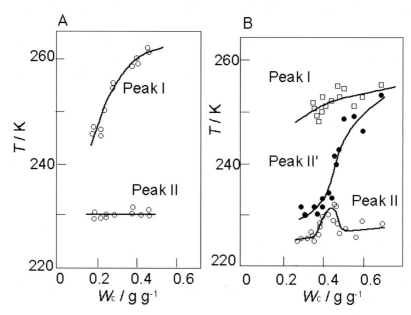

Figure 3-25. Relationship between peak temperature of crystallization and water content, (A): cellulose I, (B): cellulose II [47].

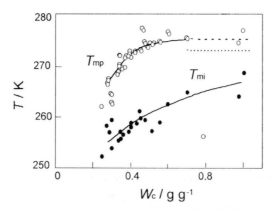

Figure 3-26. Peak temperature and starting temperature of melting of water restrained by regenerated cellulose. T_{mi}: starting temperature of melting, T_{mp}: peak temperature of melting, broken line indicates 273 K, Experimental conditions; see the caption of Figure 3-24 [47].

3.2 Heat capacity of cellulose in the presence of water

3.2.1 X-ray diffractogram of cellulose in the presence of water

As described in 3.1, crystallinity (x_c) of natural and regenerated cellulose in the dry state ranges from 0.3 to 0.8. Table 3-7 shows amorphous content ($1 - x_c$) of various kinds of cellulose in dry state estimated by x-ray diffractometry.

Table 3-7. Amorphous content of various kinds of cellulose estimated by x-ray diffractometry

Cellulose		Amorphous content A*	Amorphous content B**
Natural	bleached cotton linter	0.30	0.31
	purified jute	0.33	0.34
	purified ramie	0.38	0.30
	soft wood pulp	0.46	0.30
Regenerated	polynosic rayon	0.55	0.56
	high tenacity rayon	0.66	0.64
	viscose rayon	0.62	0.62

A* Herman's method modified by Watanabe (Watanabe, S. and Akabori, T., J. Ind. Chem. Japan, 72, 1565 (1969)

B* crystallinity index

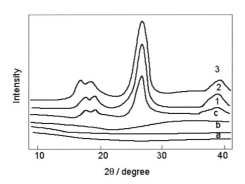

Figure 3-27. X-ray diffractogram of natural cellulose in dry and wet state. 1: dry sample, 2: wet sample (RH = 35 %), 3: wet sample (RH = 65 %), a: sample cell, b: sample cell with air with 100 % RH, c: sample cell + water, Sample preparation; see 2.5.1.

It is thought that water molecules diffuse into the amorphous region but not into the completely crystalline region of cellulose. At the same time, it is also known that molecular arrangement of amorphous chains changes when water content gradually increases. The transient process can clearly be observed when a wide line x-ray diffractogram of natural cellulose (cellulose I) is taken in the presence of various amounts of water, especially in the low water content range. Figure 3-27 shows wide-line x-ray diffractograms of natural cellulose (cotton lint). When the half width of (002) plane is plotted against W_c, the half width markedly decreases as shown in Figure 3-28. The above change is reversible, i.e. the half width expands when the natural cellulose is gradually dried.

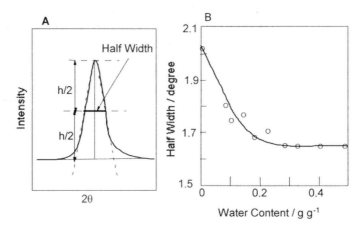

Figure 3-28. Half width (A) of (002) plane of natural cellulose (cotton lint) as a function of water content (B) [49].

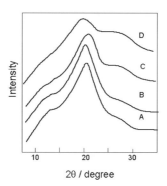

Figure 3-29. X-ray diffractograms of regenerated cellulose with various water contents. A: original sample, B: $W_c = 0.25$, C: $W_c = 0.73$, D: $W_c = 1.95$ g g^{-1}. Experimental conditions; Sample preparation; The cellulose samples were ground to a fine powder to eliminate any effect of fibre orientation. The powder was packed in a hole made in an acrylic plate with 2 mm thickness which was covered with 7 μm aluminium foil on one side using epoxy resin adhesive. The other side of the hole was sealed by the foil using silicone grease. After a determined amount of water was added to the cellulose sample, the plate was sealed with the foil and weighed. After x-ray measurements, the foil was taken off to allow water to evaporate in a heated air oven. X-ray diffractometry; A Rigaku Denki Co. x-ray diffractometer, 35 kV 20 mA Cu Kα radiation, 2θ = 3 to 35 degrees [39].

Figure 3-29 shows x-ray diffractograms of regenerated cellulose with various W_c's. No differences in the diffractograms were observed for samples having a W_c lower than 0.25 g g^{-1}. In the diffractogram of W_c = 0.73, (002), (101), and $\overline{(101)}$ peaks decreased suggesting the decrease of crystallinity of cellulose II. The shoulders observed from 26 to 28 degrees in the samples with W_c 0.74 and 1.95 g g^{-1} coincide with those of bulk water. The above facts indicate that samples C and D contain free water. The shoulder peak was not observed for samples having only bound water.

3.2.2 Mechanical properties of cellulose in the presence of water

It is known that natural cellulose shows high breaking strength in wet state. Figure 3-30 shows the relationships between relative breaking strength [= (σ_b / σ_0), where σ_0 is breaking strength of completely dry cellulose and σ_1 is that of cellulose with W_c = 0.1 g g^{-1}] and relative elongation at break [=($l_{0.1}$ / l_0), where l_0 is elongation at break of completely dry cellulose and $l_{0.1}$ is that of cellulose with W_c = 0.1 g g^{-1}]. As clearly seen, breaking strength of natural cellulose increases [49] and in contrast that of regenerated cellulose decreases.

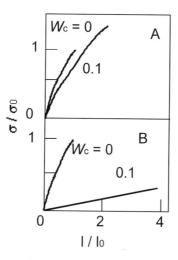

Figure 3-30. Relative stress-strain curves of natural (A) and regenerated cellulose (B). σ_0: breaking strength of completely dry cellulose, σ_1: cellulose with W_c= 0.1 g g^{-1}, l_0: elongation at break of completely dry cellulose, l: elongation at break of cellulose sample with W_c = 0.1 g g^{-1} [39, 49].

Figure 3-31 shows the relationship between relative breaking strength [= (σ_b / σ_0), where σ_0 is breaking strength of completely dry cellulose and σ_b is that of cellulose with various W_c's] and water content. The results shown in Figures 3-30 and 3-31 strongly suggest that mechanical strength of natural cellulose increases in a characteristic amount of water. When W_c exceeds ca. 0.2 g g^{-1}, mechanical strength maintains a constant value. Moreover, the above phenomena are reversible. It is known that molecular packing of biopolymers, such as chitosan, collagen and natural cellulose takes a more ordered structure in the presence of water. Mechanical strength of regenerated cellulose markedly decreases and elongation increases in the presence of water. The x-ray results shown in Figures 3-27, 3-28 and 3-29 also support the results obtained by mechanical tests.

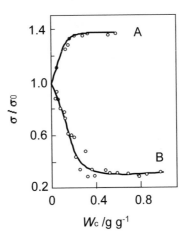

Figure 3-31. Relationship between relative breaking strength and water content of cellulose. A: natural cellulose, B: regenerated cellulose.

3.2.3   Heat capacities of cellulose

Heat capacities of cellulose of natural and regenerated cellulose are shown as a function of water content in Figure 3-32. C_p values of natural cellulose decrease at around $W_c = 0.05$ g g^{-1} and show a minimum at around $W_c = 0.9$ g g^{-1}, indicating that the higher order structure of natural cellulose is stabilized in the presence of a small amount of water. If it is assumed that C_p values of cellulose water system obey a simple additivity law, the values should be on the dotted line in Figure 3-32.

$$C_p = C_{pwater}W_c + C_{pcellulose}(1 - W_c) \tag{3.4}$$

where $C_{p\,water}$ is C_p of water and $C_{p\,cellulose}$ is C_p of cellulose. In the case of regenerated cellulose, exprerimentally obtained C_p values coincide well with calculated values, as shown in line B in the figure [52]. Since the higher order structure of natural cellulose and the structure of water vary as a function of W_c, $C_{p\ water}$ and $C_{p\ cellulose}$ changes as a function of W_c. Accordingly, the equation is not applicable for natural cellulose - water systems. The structure stabilization of natural polymers in the presence of water is known not only for cellulose but also collagen, lysozyme, etc. Discussion on C_p of polymers in the presence of water is found elsewhere [53, 54].

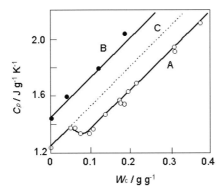

Figure 3-32. Relationship between C_p and W_c. A: natural cellulose, B: regenerated cellulose, C: calculated value.

3.3 Bound water restrained by cellulose

As already shown in Figure 3-24, the first-order phase transition is not detected until a critical amount of water is added to a polymer. In the schematic DSC cooling curves of water shown in Figure 3-33, no transition is observed as shown in curve A. The amount of water showing no phase transition is defined as non-freezing water content (W_{nf}). The maximum W_{nf} depends on hydrophylicity of polymers. In the case of cellulose, the numbers of hydroxyl group in the amorphous region, defects of crystallite and the surface of crystallite affect the W_{nf}. When W_c in the polymers exceeds a critical amount (the maximum amount of W_{nf}), a small peak (peak II) is observed at a temperature lower than the crystallization peak of bulk water (curve B). This amount of water is categorized as freezing bound water content (W_{fb}). Free water (W_f) is shown as peak I in curves C and D. W_f is unbound water content in polymers whose transition temperature and enthalpy are equal to those of bulk water (curve D).

$$W_c = W_{nf} + W_{fb} + W_f \text{ (g g}^{-1}\text{)} \tag{3.5}$$

From the cooling cycle data, the proportion of the amount of free water (W_f) is calculated by dividing the total area of the freezing water peak (Peak I) by the heat of crystallization of bulk water. The heat of crystallization is not constant for all water fractions, therefore, W_{fb} cannot be determined in the same way. It is considered that Peak II represents the irregular structure of ice formed under the influence of the hydroxyl group of cellulose molecules. Ordinarily, the amount of W_{fb} is small compared with W_{nf} (g g^{-1}) and W_f (g g^{-1}). On this account, the total area of the $W_f + W_{fb}$ (g g^{-1}) (peak I + II) per gram of dry sample is plotted as a function of W_c g g^{-1}). The intercept of the linear plot is adopted as the amount of W_{nf} (g g^{-1}).

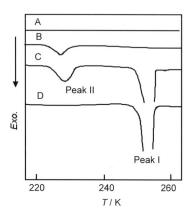

Figure 3-33. Schematic DSC cooling curves of water restrained by natural cellulose, Heating rate = 10 K min^{-1}, A: non-freezing water (W_{nf}), B: W_{nf}: freezing bound water (W_{fb}), C: $W_{nf} + W_{fb}$ + free water (W_f), D: bulk water.

The relationships between W_f, W_{fb} and W_c of natural and regenerated cellulose are shown in Figure 3-34 (A). Free water can be observed above 0.20 g g^{-1} for natural cellulose and 0.40 g g^{-1} for regenerated cellulose, and the amount increases linearly with increasing W_c. The amount of W_{fb} of regenerated cellulose calculated from the enthalpy of summation of Peak II and II' shown in Figure 3-34 (B) increases at $W_c = 0.25$ g g^{-1} and attains a maximum at $W_c = 0.40$ g g^{-1}, then decreases and levels off when W_c exceeds 0.50, where Peak II' (see Figure 3-24) merges into Peak I. The maximum point agrees well with the W_c where free water appears. When compared with W_{fb} of cellulose I, the above mentioned maximum point can only be observed in regenerated cellulose. This suggests that the amorphous region of cellulose changes gradually to a more random arrangement with increasing amount of water. This is supported by the fact that a large amount of amorphous region (see Table 3-1) and change of mechanical properties were observed.

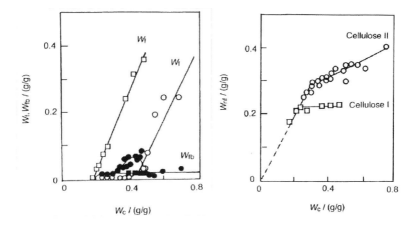

Figure 3-34. (A) Relationship between free water content (W_f), freezing bound water content (W_{fb}) and water content (W_c) of natural and regenerated cellulose. (B) Relationship between non-freezing water content (W_{nf}) and W_c of natural and regenerated cellulose I: cellulose I, II: cellulose II, Experimental conditions, see Figure 3-24.

Figure 3-34 (B) shows the relationship between W_{nf} and W_c of celluloses I and II. The W_{nf} value for cellulose I is almost constant when W_c is over ca. 0.2 (g g^{-1}), while that for cellulose II gradually increases, with increasing W_c. The above facts suggest that the amorphous region of cellulose I takes a more ordered structure with a small amount of water and this structure is stable, even with the sorption of more water. However, the amorphous region of cellulose II is not stabilized by forming an ordered structure, but increases gradually with the further sorption of more water as shown in Figure 3-34. The above results were also supported by the x-ray diffraction data. The x-ray diffractograms of cellulose I showed that the peak reflecting (002) plane became more pronounced with increasing W_c, showing the increase of the ordered structure with the sorption of water. On the other hand, the peak reflecting (002) plane of cellulose II decreased and became flatter with increasing W_c, showing the crystallinity of cellulose II. It was observed that the breaking strength of cellulose I increased with sorption of a small amount of water.

Crystallinity of natural cellulose varies according to the plant species. When purification methods are similar, it is possible to obtain cellulose I samples having various crystallinities. Bound water content of various kinds of natural polymers was quantified by DSC. Assuming the bound water attaches to the hydroxyl groups in the amorphous region, the number of water molecules restrained by each glucose unit of cellulose was calculated. Amorphous content was determined using x-ray data. Figure 3-35 shows the

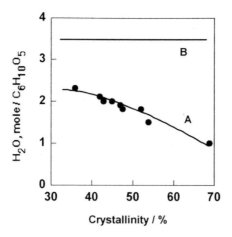

Figure 3-35. Relationship between number of bound water molecules attached to one glucose unit of cellulose and crystallinity, A: original data of bound water, B: calculated values assuming that bound water molecules attached to the hydroxyl groups of amorphous region of cellulose.

number of bound water molecules attaching to one glucose unit (curve A) and calculated values assuming that bound water molecules are restrained by hydroxyl groups in the amorphous region of cellulose (line B). The calculated values were maintained at ca. 3-4 regardless of crystallinity. One glucose unit of cellulose has three hydroxyl groups. The number of bound water molecules for each hydroxyl group is ca. 1.1.

3.4   Vaporization of non-freezing water from cellulose

Dryness of hydrophilic polymers is difficult to measure due to the fact that strong hydrogen bondings are established between the hydrophilic groups of cellulose and water molecules. In order to confirm whether water remains in the sample or not, water vaporization is measured by TG and DTA, since a trace amount of water can easily be detected due to large amounts of vaporization heat. TG and DTA vaporization curves are markedly affected by various measurement factors, such as sample mass, shape (surface area), shape of crucible, heating rate, flow rate of atmospheric gas, etc. Accordingly, it is necessary to maintain identical experimental conditions in order to obtain reliable results [55, 56].

Figure 3-36 shows water vaporization curves from cellulose diacetate measured by heat-flux type DSC. When water is cooled from room temperature to 170 K at a cooling rate of 10 K min^{-1}, crystallization was observed at 255 K due to super cooling. In the heating curve, vaporization

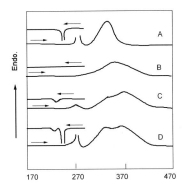

Figure 3-36. Schematic DSC curves of water vaporization curves of cellulose. A: bulk water, B: non-freezing water, C: non-freezing water + freezing bound water, D: non-freezing water + freezing bound water + free water. Measurements; heat-flux type DSC (Seiko Instruments), open type aluminium pan without lid was used. Scanning rate = 10 K min[-1]. Without gas flow.

starts immediately after completion of melting. As shown in curve A in the figure, vaporization is completed at a temperature lower than 373 K and peak temperature is observed at around 350 K. Curve B shows a schematic vaporization curve of non-freezing water. Vaporization is completed at around 450 K, suggesting that water molecules are strongly restrained by hydrophilic groups of cellulose diacetate. Curve C shows a vaporization curve of non-freezing water and freezing bound water, and curve D shows three kinds of water, non-freezing, freezing bound and free water. The results shown in Figure 3-36 indicate that bound water content can be quantitatively estimated using vaporization curves when the experimental conditions are defined.

Compared with other methods, data obtained by vaporization measurements is markedly affected by experimental conditions. Reproducibility of the experiment is not high when it is compared with the method using melting or crystallization curves using DSC. However, it is possible to obtain further information when vaporization curves are measured systematically. For example, Figure 3-37 shows vaporization curves of various amounts of non-freezing water restrained by natural (A) and regenerated cellulose (B). Water content in the figure shows no first order phase transition by DSC. Water molecules shown in this figure are restrained hydroxyl groups in the amorphous region and no-free water exists in the system. Vaporization curves show two or three peaks, especially when the water content is low. Vaporization curves of water from natural cellulose vary in a more complex manner than those of regenerated cellulose, suggesting the amorphous structure of natural cellulose is inhomogeneous.

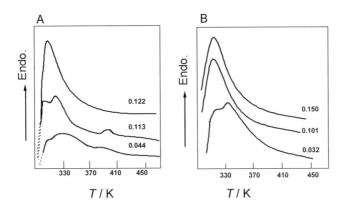

Figure 3-37. DSC curves of vaporization of non-freezing water restrained by cellulose. Numerals in the figures show water content, A: cellulose I (cotton), B: cellulose II (viscose rayon, Experimental conditions; heat-flux type DSC (Seiko Instruments) heating rate = 10 K min^{-1}, sample mass 1.6 - 1.9 mg (cotton), 1.2 -1.5 mg (viscose rayon) [56].

Hydrophilic samples such as cellulose adsorb water vapour in air during sample handling.

Isothermal vaporization of non-freezing water restrained by natural cellulose was also carried out. Figure 3-38 shows an isothermal vaporization curve at 323 K where major vaporization occurs within 20 seconds and terminates at around 100 seconds. As described above, vaporization is markedly affected by experimental conditions, especially size and mass of sample and flow rate of atmospheric gas. The details are found elsewhere [56].

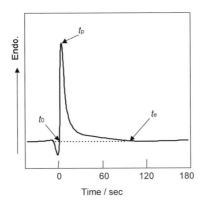

Figure 3-38. Isothermal vaporization curve of non-freezing water restrained by cellulose I (cotton). W_c = 0.0044 g g^{-1}, N_2 flow rate = 30 ml min^{-1}, temperature = 323 K.

By TG, the amount of water restrained by green polymers is also quantitatively obtained, although the initial condition of sample handling affects the data. The amounts of bound water ($W_{nf} + W_{fb}$) obtained by vaporization method measured by TG and DTA were compared with those obtained by DSC, although samples used were not cellulosic materials but lignin model compounds. Results accorded with each other in a certain range, i.e, 5 to 15 % difference was observed when W_c was larger than 0.1 g g^{-1}, however when the bound water content is smaller than 0.05 g g^{-1}, the amounts of bound water measured by vaporization method were smaller than those measured by DSC.

3.5   Visoelasticity of cellulose in water

3.5.1   Viscoelastic measurements of cellulose in humid conditions

In order to study the effect of water on viscoelastic properties of green polymers, in the initial stage of investigation, samples sorbing a certain amount of water were measured using a standard apparatus from 120 to ca. 400 K without any special equipment. Sorbed water evaporated immediately after melting of water. On this account, relaxation phenomena at a temperature lower than 273 K were reliable. In the last 20 years, various types of handmade [57] and commercially available apparatuses by which temperature and relative humidity can be controlled, have been developed (ref. Chapter 2, 1.4). Two methods are ordinarily used, (1) atmosphere of the sample cell is purged with moisture with known relative humidity, and relative humidity changes stepwise at a constant temperature, (2) the sample is immersed in water and temperature of water controlled gradually [58-62].

Figure 3-39 shows the dynamic viscoelasticity of regenerated cellulose as a function of relative humidity (RH) at 303, 323 and 354 K. Cellophane films were used. When the sorption isotherm of cellophane was measured as a function of RH at various temperatures, typical exothermic behaviour was observed. The water regain at RH 100 % was ca. 30 % at 303 K, 27 % at 323 K and 25 % at 353 K. The amount of water regain decreased with increasing temperature in the whole RH range. As shown in Figure 3-39, the dynamic modulus E' decreased with increasing RH. The above facts are explained as a scission of hydrogen bonding and plasticization of amorphous region by absorbed water. At 303 K, E' decreases slightly even at a high RH region. It is thought that the cellophane sample is in a glassy state in the whole RH range at 303 K. In contrast, E' decreased markedly at 60 - 70 % at 323 and 353 K. Tan δ peak was observed at 95 % RH at 323 K and 90 % RH at 353K. From sorption isotherms, it was confirmed that tan δ peak corresponds to 0.30 g g^{-1} at 323 K and 0.23 g g^{-1} at 353 K. By DSC, free

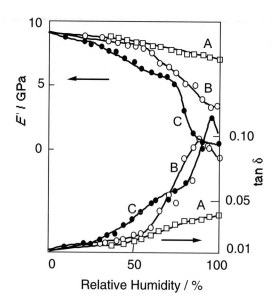

Figure 3-39. Relationship between dynamic modulus (*E'*), tan δ and relative humidity of regenerated cellulose. A: 303 K, B: 323 K, C: 353 K, Sample; Cellophane film without additives, thickness = 18 μm, uniaxial oriented, birefringence index S_n = 8.6 x 10^{-4}, crystallinity = ca. 48 % (by x-ray diffractometry) Measurements; dynamic viscoelastic measurements, Rheovibron DDV-IIC (Toyo Baldwin) equipped with a moisture generator, frequency = 110 Hz.

water is observed at W_c = 0.23 g g^{-1}. The above facts suggest that the drastic decrement of *E'* and tan δ peak is related to the formation of free water. It is considered that molecular enhancement of amorphous chains is observed in the presence of water. The longitudinal and transverse relaxation times (T_1 and T_2) were measured as a function of water content at 298 K by nuclear magnetic relaxation measurements in a W_c from 0.2 to 0.35 g g^{-1}. T_1 decreased from 1.0 to ca. 0.4 sec and T_2 increased from 0.1 to 3.5 msec, indicating water molecules restrained by cellophane films are in the state of a non-rigid solid. X-ray diffractometry was carried out during water sorption (see Figure 2-20) and it was found that the half width of the peak and intensity were varied at RH where *E'* decrement was observed.

Figure 3-40 shows *E'* and tan δ curves of untreated cellophane film in water from 274 to 268 K. Both heating and cooling curves are shown. A large hysteresis is observed, i. e., on heating, *E'* values decreased from 0.7 GPa to 0.4 GPa, and on cooling, *E'* increased from 0.4 GPa to 1.15 GPa. Tan δ peak is observed at 283 K on heating and 293 K on cooling. When cellophane film is treated in boiling water for 48 hrs, tan δ peak was observed at the same temperature, although the hysteresis found in *E'* curve decreased. *E'* values was measured isothermally at 343 K as a function of time and activation energy was calculated as 28 kJ mol^{-1}, assuming the Arrhenius type relaxation. It is thought that the relaxation at around room temperature is the local mode relaxation attributable to the cooperative motion of the pyranose ring and water.

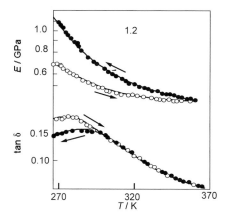

Figure 3-40. Dynamic modulus (*E'*) and tan δ of untreated cellophane in water. Sample; see caption of Figure 3-39, Measurements; Rheovibron DDV-IIC (Toyo Baldwin) equipped with a moisture generator, frequency = 110 Hz, heating and cooling rate = 1 K min^{-1}, Temperature range 274 -268 K.

3.6   Structural change of water in cellulosic hollow fibres

When cellulosic materials are used as membranes, in biomedical materials, etc., it is important to investigate the water-cellulose interaction [63-65]. When the transport properties of water or small molecules in cellulosic materials are investigated, not only the role of non-freezing water, but also the role of freezing water having disordered structure is thought to be important. In this section, phase transition behaviour of water trapped in cellulosic hollow fibres is described. Among polymeric membranes, hydrophobic synthetic polymers are also used for filtering contaminants from water. In the above polymers, the size of pores in the membranes is calculated by using super-cooling of crystallization of water. However, the property of water in the hollow fibres obtained from hydrophilic polymers seems to behave in a more complex manner.

3.6.1   Preparation of cellulosic hollow fibres

Cellulose triacetate powder (CTA) (acetyl contents, 39.8, falling ball viscosity; 10 sec) was dissolved in 1-methyl-2-pyrrolidinone and kept for 12 hours at room temperature. The concentration was 9, 12, 15, 18 and 20 %. A fibre spinning system is shown in Figure 3-41. Dope (CTA solution) was directly introduced into the water bath through a spinlet. Water was continuously supplied to the winding machine so as not to dry up the fibre. Immediately after winding, the fibre was washed by flowing water in order to eliminate the solvent. Part of the fibre was immersed in NaOH-water-ethanol solution and hydrolyzed for 12 hours at room temperature. The hydrolyzed fibre was repeatedly washed in water until it became neutralized. At the same time, cellulose films were prepared from cellulose triacetate films using a method similar to that described above. By infrared spectroscopy, it was confirmed that the absorption band of the acetoxyl group disappeared after hydrolyzation. Never-dried cellulose acetate and cellulose fibres were used for the measurements. A cross section of hollow fibres is shown in Figure 3-42.

Size of fibre (denier and radius) and water content depend on the concentration of dope. When the dope concentration (wt %) increased from 9 to 20 %, water content [(mass of hollow fibre as spun) / (mass of dried fibre), g g^{-1}] decreases from 15 to 5 g g^{-1}, and the radius of fibre increased from 0.15 to 0.31 mm. Cross section of CTA hollow fibres measured by scanning electron microscopy is shown in Figure 3-42.

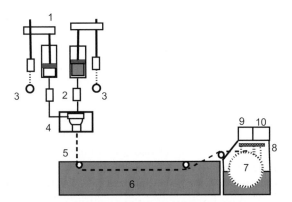

Figure 3-41. Preparation of cellulosic hollow fibre using a spinning apparatus. 1: syringe type pump, 2: filter, 3: speed controller, 4: spinlet (size of dope nozzle = 1.5 mm, diameter of nozzle = 0.8 mm, core solvent nozzle = 0.5 mm), 5: hollow fibre sample, 6: coagulating bath, 7: fibre winder, 8: water shower, 9: winding counter, 10: winding speed controller, distance from nozzle to the surface of coagulating bath = 8 cm [66].

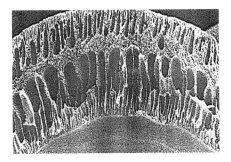

Figure 3-42. Scanning electron micrograph of cross section view of cellulosic hollow fibre.

3.6.2   Melting of water in hollow fibres

Figure 3-43 shows DSC melting curves of cellulose triacetate (CTA) and cellulose hollow fibres with various water contents. The low temperature side peak is characteristically seen for both samples. When chemical structure is taken into consideration, in the case of CTA hollow fibres, water molecules interact with acetate groups via weak hydrogen bonding. This

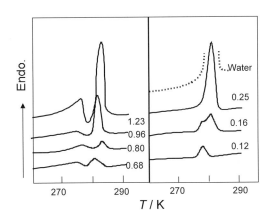

Figure 3-43. DSC heating curves of water restrained by cellulose triacetate and cellulose hollow fibres with various water contents. A: cellulose triacetate hollow fibre, B: cellulose hollow fibre. Experimental conditions; Sample preparation, CTA; Kodak CA-398-10, degree of acetylation 39.8 %, viscosity index 10, dope concentration 15 wt %. Solvent; 1-methyl-2-pyrrolidinone Water content of the never-dried sample was changed by gradual evaporation. Power compensation type DSC (Perkin Elmer), heating rate = 10 K min^{-1}, sample mass = 2 - 3 mg. Calibration material; pure water.

suggests that the bound water content of CTA hollow fibre is small and most water in the fibres is freezing water (= free water + freezing bound water).

The melting of water in CTA hollow fibres starts at 265 K and shifts slightly with increasing W_c. In contrast, the low temperature side melting of cellulose hollow fibre starts to deviate from the baseline at around 255 K. The main peak temperature of cellulosic hollow fibre is 5 K higher than that of CTA fibre. Water-hydroxyl group interaction is thought to be established in cellulose hollow fibres.

3.6.3   Pore size distribution of cellulose triacetate hollow fibre

Pore size of hollow fibre can be controlled by changing spinning conditions, such as dope concentration, conformation of spinlet, winding speed and coagulation solvent. Pore size is also controlled by adding a controlling agent in the dope. For example, poly(vinylpyrolidone) (PVP) ($M_w = 1.0 \times 10^4$) was used as a controlling agent and was mixed with CTA using 1-methyl-2-pyrrolidinone as a solvent at 313 K. In order to eliminate PVP, 15 % methanol was added to water as a core solvent. Figure 3-44 shows a flow diagram of sample preparation and measurements. For reference, flat membrane was prepared using a similar method on a glass plate. Water content of CA hollow fibres spun without PVP ranged from 4.8 to 5.4 g g^{-1} (standard deviation +/- 0.5 g g^{-1}) and fibres spun with PVP were from 6.1 to 6.5 g g^{-1}. It is thought that residual PVP in hollow fibres is negligible.

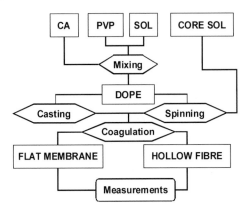

Figure 3-44. Flow diagram of sample preparation and measurements.

Pore size distribution of CA hollow fibres and surface morphology observed by scanning electron microscope is shown in Figure 3-45. By adding pore controlling agent (PVP), pore size distribution broadens. It is also clearly seen that the distribution is affected by spinning conditions.

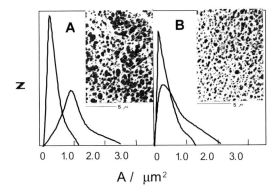

Figure 3-45. Pore size distribution curves of CA hollow fibres and representative SEM micrographs of cross section. A: spun without pore size controlling agent (PVP), B: with PVP, I: core solvent water, II: 15 % methanol aqueous solution, Pore size was analyzed using scanning electron micrographs by image analysis software.

Figure 3-46 shows DSC heating curves of the water restrained by never-dried CA follow fibres prepared by different conditions. Broad melting peaks with low temperature side peak are observed. The low temperature side peak is attributed to the freezing bound water which is restrained by the network matrix. Compared with DSC curves of hydrophobic hollow fibres, such as poly(methyl methacrylate), clear separation of both peaks was not observed for CA hollow fibres. This is due to hydrophylicitiy of CA molecular chains.

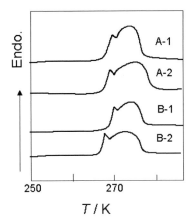

Figure 3-46. DSC heating curves of water restrained by CA hollow fibres. A: spun without pore size controlling agent (PVP), B: with PVP, 1: core solvent water, 2: 15 % methanol aqueous solution, Experimental conditions; power compensation type DSC (Perkin Elmer), heating rate = 10 K min^{-1}; sample mass = ca. 4 mg, alminium sealed type pan was used, N$_2$ flow rate = 30 ml min^{-1}.

3.7   Crystallization of amorphous cellulose in the presence of water

As described in 1.4 in this chapter, amorphous cellulose prepared by saponification of cellulose acetate crystallizes in the presence of water. Figure 3-47 shows DSC heating curves of amorphous cellulose treated at various relative humidities at room temperature. Broad exotherms show the formation of inter-molecular hydrogen bonds. Enthalpy of exotherm decreases with increasing RH %. The amount of residual free OH groups can be estimated from the area of exotherm [29, 67, 68].

Figure 3-48 shows wide angle x-ray diffractograms of various kinds of amorphous cellulose, treated amorphous cellulose and regenerated cellulose. It is clearly seen that amorphous cellulose crystallized to cellulose II type crystal when it is kept under humid conditions.

Figure 3-49 shows relative values of non-reacted residual free OH groups of amorphous cellulose maintained at 80 and 100 % RH as a function of time. After a certain induction time, the enthalpy decreases gradually. Relative value of non-reacted OH groups to total amount is calculated as,

$$\alpha = \frac{\Delta H_t}{\Delta H_0} \tag{3.6}$$

where ΔH_t is enthalpy of exothermic peak of the sample treated for t min and ΔH_0 is that of original sample without any treatment.

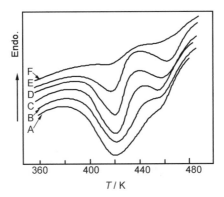

Figure 3-47. DSC heating curves of amorphous cellulose treated at 80 % for various times, Numerals in the figure show treating time (min). Sample preparation, see secton 2.4. of this chapter. Measurements; power compensation type DSC (Perkin Elmer), heating rate = 10 K min⁻¹ N₂ flow atmosphere, sample mass = ca. 8 mg.

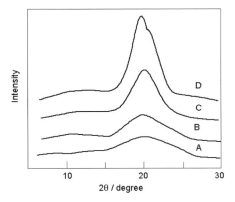

Figure 3-48. X-ray diffractrogrms of amorphous cellulose (A: amorphous cellulose annealing at 383K for 4 hrs in completely dry state (B), recrystallized cellulose by immersing amorphous cellulose in water for 24 hrs (C) and regenerated cellulose (cellophane) (D).

The decrease of α values suggests that the hydrogen bonding and/or crystallization occurs during treatment under humid conditions. In order to confirm the above fact, the rate of H $\rightarrow$ D exchange was measured by IR spectrometry, since absorption bands, OH and OD stretching, have a relation to amorphous fraction of cellulose. The fraction of OH groups which are located in the amorphous regions was measured by deuteration of the samples in the vapour phase (Sample cell designed for H to D exchange was shown in Figure 2.19).

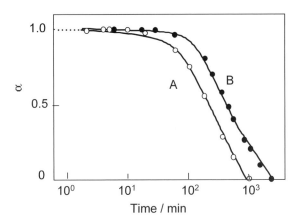

Figure 3-49. Relationships between the relative fraction of non-reacted portion and time of conditioning of amorphous cellulose, A: RH = 80, B: 100 %.

Figure 3-50 shows representative IR spectra of amorphous cellulose that contacted with D_2O vapour for various time intervals. The baseline optical density (ref. Chapter 2 Figure 2-20) of the OH band at 3360 cm^{-1} and OD band at 2530 cm^{-1} were calculated using the band of 2900 cm^{-1} (CH and CH_2 stretching) as an internal standard. Using an internal standard, differences among samples such as film thickness can be cancelled out. The ratio of the baseline optical density of the two bands is given by

$$\text{Optical density (OD)} = \log\left(\frac{I_0}{I}\right)OD/\log\left(\frac{I_0}{I}\right)CH \tag{3.7}$$

I_0 corresponds to line AB and I to line CB shown in Chapter 2 (Figure 2.20) Relative optical density (ROD) is defined as follows,

$$ROD = \frac{OD_t}{OD_\infty} \tag{3.8}$$

where OD_t is the relative optical density of the OD band of the sample treated for a certain time t and OD_∞ is the relative optical density of OD band at the time where no further reaction can be observed. Figure 3-51 shows the representative relationship between ROD and deutration time. Each sample was treated by H_2O vapour for a predetermined time. After the sample was completely dried, D_2O vapour was introduced to the sample and maintained until IR spectra maintained a constant pattern. From the curves shown in Figure 3-51, it takes time for H D exchange when the sample is treated in H_2O vapour shorter than 10^2 min. However, treating time in H_2O vapour increased, H m D exchange is carried out more rapidly.

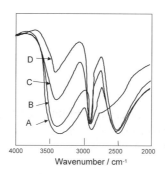

Figure 3-50. Representative infrared spectra of amorphous cellulose films deuterated by D_2O. A: original sample, B: 100, C: 200 min, D: 300 min. Measurements; IR spectrometer (Perkin Elmer). The sample holder shown in Chapter 2 (Figure 2-19) was used. Deuteration was carried in the vapour phase. The reaction was stopped at different stages.

Using relative optical density, diffusion coefficient of D_2O to amorphous cellulose film was calculated using the following equation. For the calculation, it is assumed that the shape of sorption isotherms are not different from each other and concentration dependency of diffusion constant, D, is negligible.

$$\frac{dC}{dt} = D\left(\frac{d^2C}{dx^2}\right) \tag{3.9}$$

Where C is the concentration at time t and co-ordinate x which corresponds to thickness of cellulose film. If the film is semi-infinite, equation (1) gives on solution

$$\frac{M_t}{M_\infty} = \frac{4}{L}\left(\frac{Dt}{\pi}\right)^{\frac{1}{2}} \tag{3.10}$$

Where M_t is the amount sorbed at time t, M the amount sorbed after infinite time, and L is the thickness of the film. M_t corresponds to OD_t and M_∞ corresponds to OD_∞. Thus, equation (3) is obtained

$$\left(\frac{OD_t}{OD_\infty}\right) = \left(\frac{4}{L}\right)\left(\frac{D*t}{\pi}\right)^{\frac{1}{2}} \tag{3.11}$$

$$D^* = \left(\frac{L^2\pi}{16}\right)\left(\frac{OD_t}{OD_\infty}\right)^2\left(\frac{1}{t}\right) \tag{3.12}$$

Where D^* is diffusion constant calculated from baseline optical density. Values of OD_t / OD_∞ were plotted against $t^{1/2}$, and D^* values were calculated from the gradient of curves of samples conditioned at different relative humidity for various time intervals, Figure 3-51 shows the relationships between D^* and treatment time. D^* decreases up to ca. 10^2 min and then increases gradually.

Heat capacity of amorphous cellulose with water content 0.14 g g^{-1} was measured as a function of time. It is thought that structural change occurs while holding the sample at 325 K. As shown in Figure 3-53, the minimum point was observed at around 500 - 600 min. Non-freezing water content of amorphous cellulose depends on time of conditioning due to the change of the higher order structure. The amount of water in the sample shown in Figure 3-53 is categorized into non-freezing water (ref. 3.2.3). It is clearly seen that molecular chains are rearranged, crystalline structure is formed at around 400 min in the presence of non-freezing water and the higher order structure of amorphous regions successively varied following the increase in molecular regularity of the crystalline region.

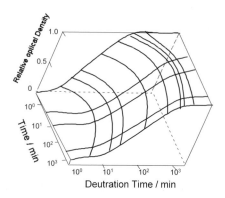

Figure 3-51. Relative optical density of amorphous cellulose maintained at RH 100 % for various times as a function of deutration time in the vapour phase.

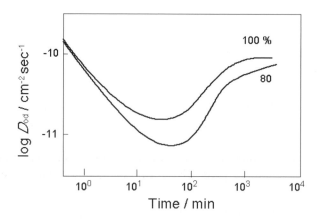

Figure 3-52. Diffusion constant calculated from change of relative baseline optical density with deueration time of amorphous cellulose conditioned at 80 and 100 % relative humidity of H_2O vapour.

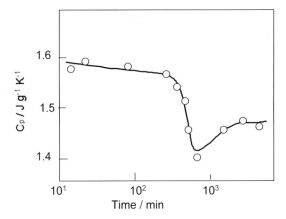

Figure 3-53. Change of heat capacities of amorphous cellulose with water content 0.14 g g^{-1} at 325 K as a function of time. Sample: amorphous cellulose was conditioned in 100 % relative humidity and sealed hermetically in an aluminium pan after 14 min conditioning. Water content at 14 min was 0.14 g g^{-1}. Heat capacity was measured by power compensate type DSC (Perkin Elmer) in a temperature range from 320 to 330 K repeatedly. The amount of water was not varied. Holding time was varied from 14 to 665 min.

Longitudinal (T_1) and transverse (T_2) relaxation times of ^{1}H relaxation were obtained for amorphous cellulose at 100 % RH at 293 K as a function of conditioning time. As shown in Figure 3-54, it is clearly seen that structure of water changes at around 10^3 minutes where C_p values also markedly changed. In the presence of small amounts of water, cellulose molecules can rearrange easily by the effect of water which causes the initial hydrogen bonding formation of cellulose. Judging from the results showing that the relative value of amorphous fraction stays constant, while conversion from H to D increases gradually in the initial stage of conditioning, the rearrangement of molecules does not occur as a whole but only partly in the vicinity of OH groups which can affect the process of H to D exchange. Then, if larger amounts of water are supplied in a wet atmosphere, the excess water breaks the initially formed hydrogen bond between OH groups, and cellulose molecules rearrange more freely and crystalline structure is formed.

Once molecular chains move in a long range order and crystalline region is formed, the molecular chains which are not included into the crystalline region are randomly arranged as an amorphous region accompanying with free volume, in which D$_2$O vapour diffuses freely. Increase of D^* in the long time conditioning can be explained by formation of the amorphous region with free space. Heat capacity data also support the above interpretation. Once molecular chains move in a long range order and crystalline region is formed, the molecular chains which are not included into the crystalline region are randomly arranged as an amorphous region accompanying with free volume, in which D_2O vapour diffuses freely. Increase of D^* in the long time conditioning can be explained by formation of the amorphous region with free space. Heat capacity data also support the above interpretation.

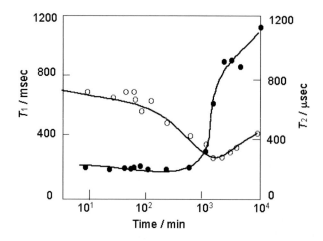

Figure 3-54. Longitudinal (T_1) and transverse (T_2) relaxation times as a function of conditioning time of amorphous cellulose at 100 % RH at 293 K. NMR spectrometry (Nicolet) T_2 values were measured by the 180-τ-90 degree pulse technique and T_2 values were obtained by the free induction decay following the 90 degree pulse.

4.   LIQUID CRYSTALS AND COMPLEXES

4.1   Phase transition of carboxymethylcellulose-water systems

In this section, phase transition behaviour of sodium carboxymethyl-cellulose (NaCMC)-water system in a W_c ranging from 0 to ca. 5 g g^{-1} (for definition of W_c, see equation 3.3) is investigated by DSC. Figure 3-55 shows stacked DSC heating curves of NaCMC with water content (W_c) ranging from 0 to 2.0 g g^{-1} in a temperature range from 150 to 350 K. All samples were cooled from 350 to 150 K at a cooling rate of 10 K min^{-1}.

From the low to high temperature side, glass transition temperature (T_g), cold crystallization temperature (T_{cc}), melting of water accompanied with shoulder peak in the low temperature side (T_d) and liquid crystal to liquid transition (T^*) are observed. When NaCMC-water system with W_c ca. 1.0 g

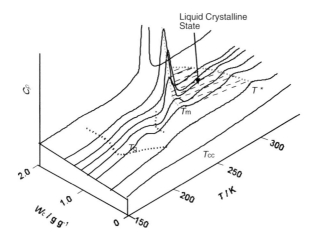

Figure 3-55. DSC heating curves of NaCMC-water systems with various water contents. T_g: glass transition temperature, T_{cc}: cold crystallization temperature, T_m: melting temperature, T^* liquid crystal to liquid transition temperature. Experimental conditions; NaCMC, degree of substitution = 0.95 Power compensation type DSC (Perkin Elmer), heating rate = 10 K min^{-1}, sample mass = ca. 3 mg, aluminium sealed type pan [69].

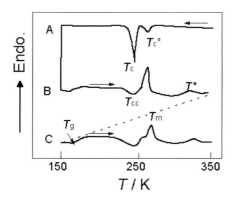

Figure 3-56. Effect of thermal history on DSC curves of NaCMC-water system with water content ca. 1.0 g g^{-1}. Scanning rate = 10 K min^{-1}, -----: quenching, Experimental conditions: see Figure 3-55 caption.

g^{-1} was observed using a polarizing microscope at 298 K, a clear liquid crystalline pattern characterized by nematic crystals was found. When W_c is less than ca. 0.4 g g^{-1}, T_g and T_c^* can be seen, T_{cc} is observed in a narrow W_c region from ca. 0.5 to 1.5 g g^{-1}, where Tm increases gradually. T^* decreases with increasing W_c. T^* could not be detected at around W_c = ca. 4.0 g g^{-1}.

Thermal history affecting the transition behaviour of NaCMC-water system with $W_c = 1.0$ g g^{-1} is shown in Figure 3-56. When the system was cooled from 350 K to 150 K at cooling rate of 10 K min^{-1}, liquid to liquid crystal transition (T_c^*) and crystallization (T_c) are reversibly observed (curve A). Glass transition, shallow exotherm due to cold crystallization, broad melting peak and liquid crystal to liquid transition can be seen in the heating curve scanned at 10 k min^{-1} (curve B). When the system is quenched, the baseline gap due to glass transition increases and cold crystallization peak increases.

Phase diagram of NaCMC water systems was established using phase transition temperatures shown in Figure 3-57. As shown in Figure 3-57, T_g decreases in the initial stage, reaches the minimum point and then gradually increases. T_{cc} starts to be detected at a W_c where T_m is also detectable, although T_m is far lower than that of bulk water. T_m gradually increases and levels off at around W_c where T_g can not be observed. T^* can be found at the W_c where T_g can be detected. This diagram indicates that molecular motion of NaCMC is enhanced when a small amount of water is introduced. Intermolecular bonding is broken when T_g is detected. At the same time, NaCMC molecules align in the same direction. When the amount of water increases, a part of frozen water crystallizes at T_{cc} into irregular ice which melts at a temperature far lower than that of bulk water. When free water is formed in the system, free molecular motion of NaCMC is restricted and T_g increases gradually. T_g becomes impossible to observe, when there is an excess amount of free water.

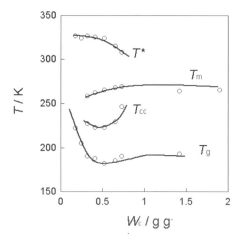

Figure 3-57. Phase diagram of NaCMC-water systems. T_g: glass transition temperature, T_{cc}: cold crystallization temperature, T_m: melting temperature, T^*: liquid crystal to liquid transition temperature.

4.2   Liquid crystal of carboxymethylcellulose substituted by various cations

CMC samples substituted by various cations as shown in Table 3-8 were measured using a similar method to that described in 3. 4. 1 in the presence of W_c = ca. 1.0 g g^{-1}. Although water solubility of substituted CMC decreased with increasing valency, molecular alignment was clearly observed for all CMC samples by polarizing light microscopy.

Table 3-8. Degree of substitution of carboxymethylcellulose with various cations

Valency	Cation	Degree of substitution
Monovalent	Li	0.61
	Na	(0.48), 0.60, (0.8, 1.4, 1.7)
	K	0.59
	Cs	0.63
Divalent	Mg	0.59
	Ca	0.57
	Ba	0.56
Trivalent	Al	0.57

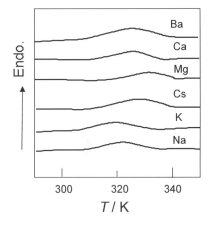

Figure 3-58. DSC heating curves of CMC substituted by various cations in the presence of water (W_c = ca. 1.0 g g^{-1}). Power compensation DSC (Perkin Elmer), heating rate = 10 K min^{-1}, sample mass = ca.6 mg [70].

By DSC measurement, a small broad peak due to phase transition from liquid crystal to liquid could be observed. In Figure 3-58, representative DSC heating curves in a temperature range of liquid crystal transition are shown. The temperature from the liquid crystal to isotropic liquid phase is observed at around 300 to 320 K and no large difference is found if counter ions are changed.

4.3   Effect of substituted mono- and divalent cations on non-freezing water content of carboxymethylcellulose-water systems

Water-cellulose derivatives interaction in the presence of cations has been investigated by many researchers [71-78]. Bound water content of NaCMC-water systems was estimated from melting enthalpy using equation 3.4. Figure 3-59 shows the relationship between non-freezing water content (W_{nf}), free water content (W_f) and W_c. The amount of W_{nf} is far larger than that of cellulose due to sodium ion. At the same time, a small peak is observed at around W_c 1.0 g g^{-1} from where freezing water can be detected. It is clear that freezing bound water of NaCMC exists in a W_c ranging from 0. 5 to 0.10 g g^{-1}, melting peak temperature was observed from W_c 0.5 g g^{-1} at a temperature far lower than that of bulk water in the phase diagram, as shown in Figure 3-59. In spite of the above facts, enthalpy of freezing bound water is almost negligible, and on this account, W_{fb} values are ignored in Figure 3-59.

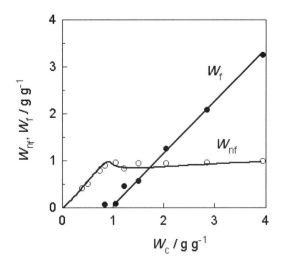

Figure 3-59. Relationship between non-freezing water content (W_{nf}), free water content (W_f) and W_c of NaCMC. DS=1.4, M_v =7.8 x 10^4.

The relationship between degree of substitution (DS) and W_{nf} is shown in Figure 3-60. W_{nf} values increase with increasing DS. This suggests that number of Na ions markedly affects the amounts of non-freezing water.

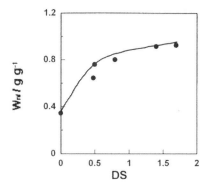

Figure 3-60. Relationship between degree of substitution (DS) and non-freezing water content of NaCMC. (M_w. See Table 3-5) [71].

When Na ions in NaCMC are converted into divalent ions, such as Ca or Fe ions in aqueous media, water insoluble CMC is obtained. The experimental procedure for sample preparation from NaCMC to CaCMC is shown in Figure 3-61 as an example. NaCMC aqueous solution was extended on a glass plate, immersed in $CaCl_2$ aqueous solution and then transferred into a water bath after coaggregation was completed. Water insoluble films or sheets were obtained after drying. The colour of films varied according to the kind of counter ions, i.e. divalent Fe was brown, Mn transparent, Sn white, Cu blue, Ca turbid white, etc. CoCMC was pink, however it was partially soluble in water. By the substitution of trivalent Al and Fe, water insoluble films were also obtained. Mechanical properties of the above obtained films were tested using an Instron type tester. Representative results of breaking strength, elongation at break and elastic modulus (E) are shown in Table 3-9. By intermolecular bonding, breaking strength increases, and in contrast elongation at break markedly decreases.

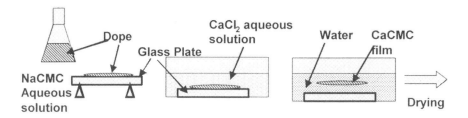

Figure 3-61. Example of sample preparation of di- and trivalent cation substituted CMC films.

Table 3-9. Mechanical properties of Na-, Cu- and FeCMC films in dry state

Ion	σ_b/ MPa	ε_b/ %	E/GPa
Na	45	19.7	1.8
Cu	59	9.0	2.6
Fe	61	3.2	2.4

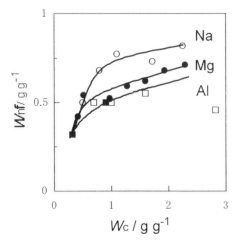

Figure 3-62. Relationship between non-freezing water content and water content (W_c) of NaCMC, MgCMC and AlCMC.

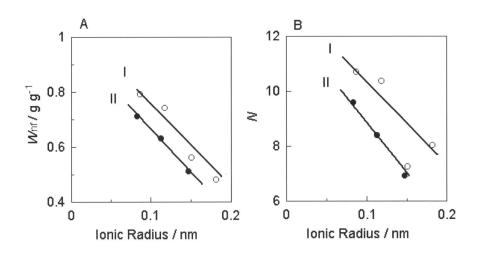

Figure 3-63. Non-freezing water content (W_{nf}) and number of water molecules attaching to each repeating unit of CMC (N) of mono- and divalent cation substituted CMC as a function of ionic radius. A: W_{nf}, B: N, I: Monovalent cations (Li, Na, K, Cs), II: Divalent cations (Mg, Ca, Ba).

Figure 3-62 shows the relationships between W_{nf} and W_c of three representative mono-, di-valent cation substituted CMC. W_{nf} value of NaCMC is the largest and W_{nf} value decreases in the order of valency. Figure 3-63 (A) shows relationships between the maximum W_{nf} (g g^{-1}) values of mono- and divalent cations substituted CMC and ionic radius (nm). W_{nf} decreases linearly with increasing W_c. The number of water molecules restrained by repeating unit of each CMC was calculated and the mol number is plotted against ionic radius in Figure 3-63(B). Concerning potassium ion, N value shown in Figure 3-63 (B) deviates from the linear line. This suggests that irregular behavior of potassium ion in aqueous solutions may be concerned with this phenomenon [72].

4.4 Heat capacity of carboxymethylcellulose in the presence of water

Figures 3-64 and 3-65 show representative C_p curves of NaCMC with various amounts of water. Water content shown in the figures is categorized into non-freezing water from the results of Figure 3-59. T_g of dry NaCMC was observed at around 410 K. Glass transition temperature (T_g) of NaCMC having non-freezing water shifts to the low temperature side.

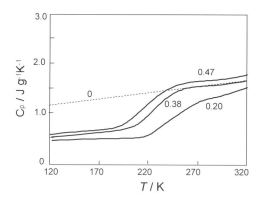

Figure 3-64. Heat capacity curves of NaCMC (DS = 1.4) in the presence of various amounts of non-freezing water. Numerals shown in the figure are non-freezing water contents (g g^{-1}). Samples; M_w see Table 3-5, NaCMC powder was solved in water (4 wt %) and extended on a glass plate. Films with 0.2 mm thickness were obtained, dried at room temperature and heated at 390 K for 10 min in order to eliminate residual water. Measurements; heat-flux type DSC (Seiko Instruments DSC 220C), open type aluminium pan, heating rate = 10 K min^{-1}, sample mass = ca. 5 mg, C_p calculation; Sapphire was used as a standard material (Seiko heat capacity software was used) [79].

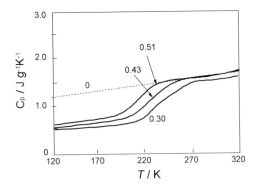

Figure 3-65. Heat capacity of NaCMC (DS = 1.7) in the presence of various amounts of non-freezing water. Numerals in the figure show non-freezing water content.

Figure 3-66 (A) shows the relationship between T_g and water content. T_g values are magnified in a W_c range from 0.2 to 0.5 g g^{-1} in Figure 3-66 (B). As shown in Figure 3-66 (A), T_g decreases abruptly in a W_c range from 0 to 0.2 g g^{-1}. As shown in the magnified figure (B), T_g decreases with increasing DS.

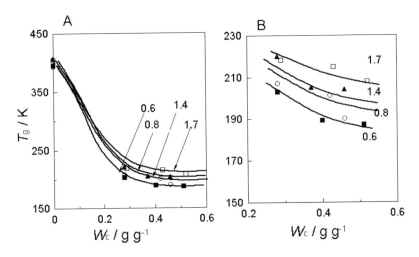

Figure 3-66. Glass transition temperature of NaCMC with various degrees of substitution and non-freezing water content (A), and magnified curves (B). Numerals in the figure show degree of substitution (DS), Experimental conditions; see Figure 3-64 caption.

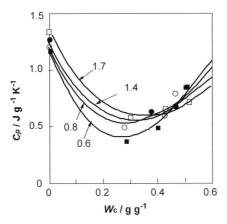

Figure 3-67. Relationships between heat capacity of NaCMC with various degrees of substitution at 173 K and water content. Numerals in the figure show degree of substitution.

Figure 3-67 shows C_p of NaCMC with various amounts of non-freezing water at 173 K. It is clearly seen that C_p values of NaCMC at the glassy state markedly decrease in the presence of non freezing water. The minimum C_p value is found at around $W_c = 0.3$ g g^{-1} regardless of M_w and DS. Hydrogen bondings and ionic bonds are dissociated by the absorption of non-freezing water and distorted molecular chains which are formed during drying reorganize gradually and more ordered structures are formed. This fact is observed as a decrease of heat capacity which suggests that CMC molecules take a more stable state in the presence of non-freezing water.

4.5   Phase transition behavior of sodium cellulose sulfate

Sodium cellulose sulfate (NaCS) is a representative cellulose ester (See Table 3-2). NaCS is a water soluble cellulose derivative [80]. NaCS-water systems form lyotropic thermotropic liquid crystal in a characteristic concentration and temperature region [81].

Figure 3-68 shows DSC curves of cellulose sulfate-water system with water content of 1.26 g g^{-1}. When the system is cooled from 350 K to 150 K at 10 K min^{-1}, a small exotherm due to transition from liquid to liquid crystalline phase and a large exotherm due to crystallization of freezing water peak are observed (curve A). In heating curve glass transition, a shallow exotherm due to cold-crystallization, an asymmetrical melting peak of water having a long tail in the slow temperature side and a broad endothermic peak due to transition from liquid crystal to liquid state are observed (curve B). A large temperature difference between (T^*- T_m) and

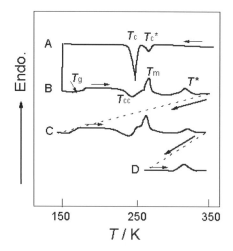

Figure 3-68. DSC curves of sodium cellulose sulfate (NaCS)-water system with water content = 1.26 g g^{-1}. A: cooling curve, B, C and D: heating curves, ------: quenching, T_g: glass transition temperature, T_{cc}: cold crystallization temperature, T_m: melting temperature, T^*: transition temperature from liquid crystalline to liquid phase, T_c^*: transition temperature from liquid to liquid crystalline phase, T_c: crystallization. Experimental conditions, Samples; degree of substitution = 2.26, Measurement; power compensation type DSC (Perkin Elmer), scanning rate = 10 K min^{-1}, sample mass = ca. 5 mg.

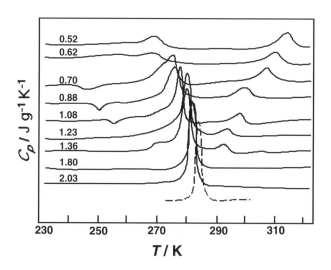

Figure 3-69. Stacked DSC heating curves of NaCS-water systems in a temperature from 230 to 320 K [81].

(T_c* - T_c) suggests that crystallization occurs immediately, once molecular alignment is completed. By quenching, baseline gap at T_g increases, cold-crystallization peak becomes larger and the sub-melting peak is observed clearly in the low temperature side of the main melting peak [82, 83].

Figure 3-69 shows stacked DSC heating curves of NaCS-water systems with various water contents in a temperature range from 230 to 350 K. All samples were heated once to 350 K and cooled to 150 K at 10 K min^{-1}. T* shifts to the low temperature side and the size of the peak decreases with increasing W_c. Cold crystallization is not clearly seen at a water content lower than 0.6 g g^{-1}. T_m gradually approaches the T_m of bulk water with increasing W_c. Glass transition was found in DSC curves of the samples with W_c range from ca. 0.3 to 1.3 g g^{-1}, although the temperature range where glass transition is observed is not shown in Figure 3-69.

Using peak temperature of glass transition, cold crystallization, melting and liquid crystalline state to liquid state transition, a phase diagram was established as shown in Figure 3-70.

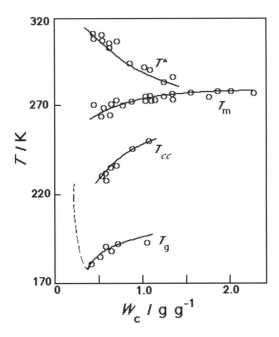

Figure 3-70. Phase diagram of sodium cellulose sulfate (NaCS)-water systems. T_g: glass transition temperature, T_{cc}: cold crystallization temperature, T_m: melting temperature, T*: transition temperature from liquid crystal to liquid state.

4.6 Heat capacity of non-freezing water restrained by sodium cellulose sulfate-water systems

The amount of non-freezing water was calculated by enthalpy of melting of NaCS-water systems. The relationship between non-freezing water content, freezing water content and water content of NaCS-water systems is shown in Figure 3-71. The maximum amount of non-freezing water is ca. 0.4 g g^{-1}.

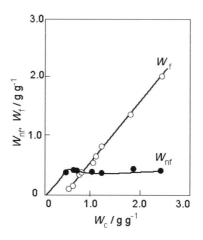

Figure 3-71. Relationship between non-freezing water content, freezing water content and water content of sodium cellulose sulfate (NaCS)-water systems. W_f: free water content, W_{nf}: non-freezing water content. Experimental conditions; see Figure 3-68 caption.

Heat capacities of NaCS-non-freezing water system (W_c = 0.15 g g^{-1}) were measured by DSC in a temperature range from 240 to 320 K. Sapphire was used as a reference material. As shown in the phase diagram in Figure 3-71, no phase transition is observed for the system with W_c = 0.15 g g^{-1} in the above temperature range. C_p values of NaCS-nonfreezing water system (Figure 3-72, curve C) in the above temperature range are larger than those calculated from simple additivity law (curve D).

If it is assumed that structural change of NaCS-non-freezing water system is attributable only to non-freezing and not to NaCS, C_p values of non-freezing water can be calculated. Calculated values are shown in Figure 3-73. C_p values are far larger than those of normal ice (hexagonal crystal). In fact, it is thought that C_p values of NaCS also decrease since molecular rearrangement commences in order to form liquid crystal. Hence, real C_p values of non-freezing water should be higher than those calculated. It can be said that non-freezing water molecules form a random and irregular structure.

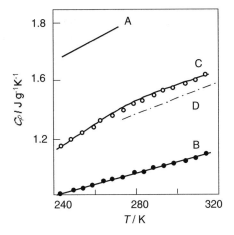

Figure 3-72. Heat capacities of dry NaCS, water and NaCS-water system with W_c 0.15 g g^{-1}. Line A: C_p of water, B: dry NaCS, C: NaCS-water system with $W_c = 0.15$ g g^{-1}, D: calculated values assuming simple additivity law. Measurement conditions. Power compensation type (Perkin Elmer), heating rate = 10 K min^{-1}, sample mass = ca. 5 mg.

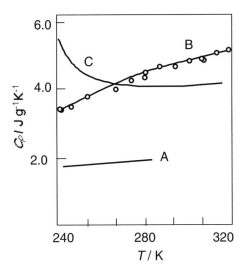

Figure 3-73. Heat capacities of water, ice and non-freezing water of NaCS-water system. A: C_p of ice, B: calculated C_p of non-freezing water, C: calculated C_p based on references [35].

4.7   Liquid crystallization of sodium cellulose sulfate-water system

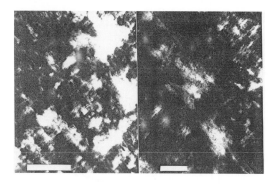

Figure 3-74. Polarized light micrograph of NaCS-water system with W_c = ca. 10 g g^{-1}. The samples were extended on a slide glass and covered with a cover glass.

Cellulose sulfate is a representative cellulose derivative in ester form (see Table 3-2). Sodium salt of cellulose (NaCS) is a water soluble polysaccharide electrolyte and is widely used in industrial fields. When highly concentrated aqueous solutions of NaCS-water systems are observed using polarized light microscopy, thermotropic lyotropic liquid crystal is

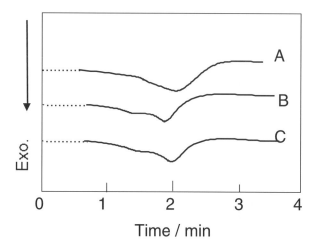

Figure 3-75. Isothermal liquid crystallization of NaCS-water systems with W_c = 0.6 g g^{-1}. 254 K (curve A), 255 K (curve B), 256 K (curve C). Experimental conditions; The samples were maintained at 320 K for 5 min, quenched to each temperature and maintained for 10 min. Power compensation type DSC (Perkin Elmer), sample mass = ca. 5 mg.

observed. Figure 3-73 shows polarized light micrographs of NaCS-water systems with W_c = ca. 10 g g^{-1}. The amount of W_c is not accurately weighed due to vaporization that occurs during measurements. When the sample was dried sandwiched between two glass plates, an oriented structure was maintained.

Isothermal liquid crystallization of NaCS-water systems was carried out by DSC at various temperatures, as shown in Figure 3-75. During isothermal liquid crystallization, two peaks were observed, especially when the temperature was high. This suggests that the liquid crystalline state is inhomogeneously formed depending on temperature. Immediately after the completion of isothermal liquid crystallization, the sample was heated at 10 K min^{-1} to 350 K. A broad endotherm with two peaks was observed.

During isothermal crystallization, the samples were annealed and the high temperature side peak was newly formed. When heating rate decreased, the high temperature peak was also observed. Enthalpy of transition increased by annealing. This indicates that liquid crystallization is accelerated by annealing. Figure 3-76 shows relationships between transition enthalpy from liquid crystalline state to liquid state (ΔH^*) and water content (W_c) of NaCS-water systems. Samples were annealed at around the transition temperature from liquid to liquid crystalline state. Enthalpies of annealed samples (line B in Figure 3-76) are higher than those of non-treated ones in a whole range of water contents.

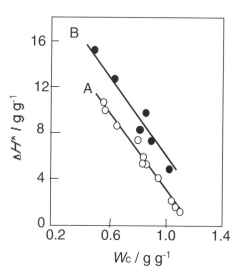

Figure 3-76. Relationships between enthalpy of liquid crystalline to liquid state transition (ΔH^*) and water content (W_c) of NaCS-water systems. A: non-treated, B: annealed. Annealing was carried out at a temperature range of T_c^* [84].

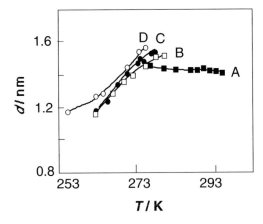

Figure 3-77. Relationships between *d* and temperature of NaCS-water systems at different W_c, A: 0.70, B: 0.89, C: 1.25, D: 1.40 g g^{-1}. Experimental conditions; Wide angle x-ray diffractometer, CuKα. 40 kV, 300 mA. Temperature was increased stepwise. Accuracy of temperature +/- 05 K [84].

In order to obtain direct evidence of the liquid crystal formation of NaCS-water system, x-ray diffractometry was carried out as a function of temperature. X-ray diffractograms of NaCS-water systems where the liquid crystal is observed by DSC, showed that 2θ = ca. 6 degrees was observed. Although x-ray measurements were carried out isothermally, the results agreed well with those obtained by DSC. In both heating and cooling processes, a broad peak at around 6 degrees was observed. Figure 3-77 shows the relationships between the intermolecular distance calculated from x-ray diffractogram (d) and temperature of NaCS-water systems with $W_c =$ 0.7 (curve A), 0.89, (curve B) 1.25 (curve C) and 1.40 (curve D) g g^{-1}. As shown in Figure 3-76, NaCS-water system with $W_c = 0.7$ g g^{-1} contains no free water, and on this account, the temperature range showing the liquid crystalline state is wide. At the same time, d values in the liquid crystalline state increase with increasing water content.

Figure 3-78 shows the relationship between the mean *d* values of the liquid crystalline state and water content. It is clear that *d* expands with increasing water content. When about two water molecules are restrained by one repeating unit of NaCS (DS = 2.2), the main chain motion is enhanced. The NaCS molecules associating with water molecules align when 4 - 5 water molecules are bound to one hydroxyl group and two Na$^+$ ions per repeating unit. The intermolecular distance calculated by x-ray diffractometry was 1.4 to 1.6 nm. The values correspond to the structure where Na$^+$ ions are sandwiched between two cellulosic chains. With increasing W_c, the intermolecular distance increases and temperature range showing the liquid crystalline state becomes narrower. The liquid crystalline phase disappears when W_c is larger than the critical value of 1.5 g g^{-1}.

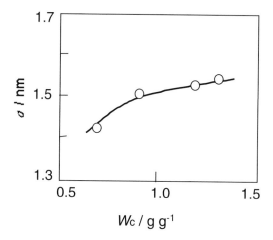

Figure 3-78. Change of intermolecular spacing (d) of liquid crystalline state of NaCS-water systems as a function of water content (W_c).

4.8 Nuclear magnetic relaxation of ^{1}H and ^{23}Na of NaCS-water as a function of temperature

The ^{1}H longitudinal relaxation time (T_1) and transverse relaxation time (T_2) of water in the NaCS-water systems with various water contents were measured as a function of temperature [69, 85-88] The T_1 and T_2 values are shown in Figure 3-79 as a function of inverse absolute temperature. A minimum value was observed in T_1 curves in a temperature range from 230 to 250 K depending on water content. The values of T_1 at the minimum point also depend on water content, and the appearance of the T_1 minimum is explained in connection with correlation time (τ_c). A marked decrease found in T_2 curves is observed in a temperature lower than 260 K, i.e. the smaller the water content, the steeper the T_2 decrease. Since the T_2 value reflects averaged molecular motion of water in the system, it is clear that molecular motion changes considerably at that temperature.

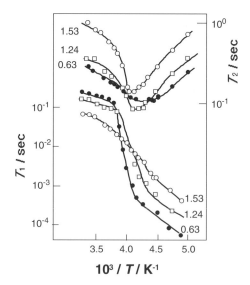

Figure 3-79. ^{1}H longitudinal, and transverse relaxation times (T_1 and T_2) as a function of reciprocal temperature of NaCS-water systems at various temperatures. Water content of samples; A: 0.83, B: 1.24, C: 1.58 g g^{-1}. Experimental conditions; T_1 was measured by the 180-t-90 degree pulse method, T_2 was measured by either the Meiboom-Gill variant of the Carr-Purcell method or the solid echo method.

Figure 3-80 shows relationships between ^{1}H longitudinal, transverse relation times (T_1 and T_2) and water content (W_c) of NaCS-water systems at various temperatures. It is only possible to distinguish a limited number of perturbed sites by NMR. On this account, in NaCS-water system, as described in the previous sections, the results of DSC indicate that two groups of water exist. If two sites represent free water fraction (P_f) and bound water fraction (P_b) then the following equations are derived.

$$P_f + P_b = 1 \qquad\qquad\qquad\qquad (3.13)$$

if it is assumed $R_i = \dfrac{1}{T_i}$ (i=1,2)

where T is relaxation time, then

$$R_i = P_f R_{if} + P_b R_{ib} \qquad\qquad\qquad\qquad (3.14)$$

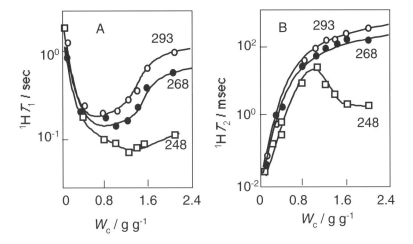

Figure 3-80. Relationships between ^{1}H longitudinal and transverse relation times (T_1 and T_2) and water content (W_c) of NaCS-water systems at various temperatures. A: 293 K, B: 268 K, C: 248 K.

In the free water, the extreme narrowing condition is fulfilled and in this case,

$$R_{1f} = R_{2f} = R_f \tag{3.15}$$

The equation 3.14 is rewritten as,

$$\frac{(R_1 - P_f R_f)}{(R_2 - P_f R_f)} = \frac{R_{1b}}{R_{2b}} \tag{3.16}$$

The relaxation rates of bound water are expressed as [89]

$$\frac{1}{T_{1b}} = G\sum_i k_i \left(\frac{\tau_{ci}}{1 + \omega_0^2 \tau_{ci}^2} + \frac{4\tau_{ci}}{1 + 4\omega_0^2 \tau_{ci}^2} \right) \tag{3.17}$$

$$\frac{1}{T_{2b}} = \frac{1}{2}G\sum_i k_i \left(3\tau_{ci} + \frac{5\tau_{ci}}{1 + \omega_0^2 \tau_{ci}^2} + \frac{2\tau_{ci}}{1 + 4\omega_0^2 \tau_{ci}^2} \right) \tag{3.18}$$

where $\sum k_i = 1$, G is the interaction constant determined by the magnitude of the relaxation, ω_0 is the angular resonance frequency and τ_{ci} is the correlation time. Although the above equations 3.17 and 3.18 assume multiple correlation times in the systems, if it is assumed that only one average correlation time τ_c represents the relaxation of bound water, the equations from 3.13 to 3.18 can be combined and equation 5.8 is obtained.

$$\frac{T_{2b}}{T_{1b}} = \frac{2\left(\dfrac{1}{1+\omega_0^2\tau_{ci}^2} + \dfrac{4}{1+4\omega_0^2\tau_{ci}^2}\right)}{\left(3+\dfrac{5}{1+\omega_0^2\tau_{ci}^2} + \dfrac{2}{1+4\omega_0^2\tau_{ci}^2}\right)} \qquad (3.19)$$

Equation 3.19 is a function of $\omega_0\tau_c$. When τ_c values calculated from the equation 3.8 are plotted against the inverse absolute temperature, the τ_c values depend on temperature not on water content. The values increased from 10^{-8} sec at 260 K to ca. 3×10^{-6} sec at 200K. The above fact indicates that the bound water in the NaCS-water system is in the state between viscous liquid and non-rigid solid in this temperature range. When the ln τ_c is plotted to the inverse absolute temperature, a linear relationship can be obtained as shown in Figure 3-81. The temperature dependency of τ_c is expressed by the Arrhenius equation as shown in the equation 3.9.

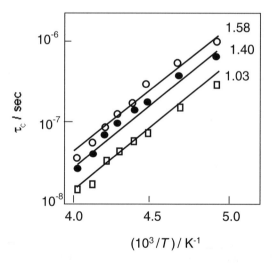

Figure 3-81. Relationships between the average correlation time (τ_c) of ^{1}H longitudinal of NaCS-water systems at various water contents. W_c, A: 1.03, B: 1.24, C: 1.40, D: 1.58 g g^{-1}.

$$\tau_c = \tau_0 \exp\left(\frac{E_a}{RT}\right) \tag{3.20}$$

The calculated E_a values were from ca. 18 to 29 kJ mol^{-1} depending on W_c.

Figure 3-82 shows the relationships between T_1 values of ^{23}Na and inverse absolute temperature (A) of representative NaCS-water systems, i.e. the sample with $W_c = 0.59$ g g^{-1} contains non-freezing water and forms liquid crystal, and the sample with $W_c = 0.96$ g g^{-1} contains non-freezing water and a small amount of free water. Since the temperature range where the liquid crystal forms is narrow, the sample with $W_c = 1.94$ g g^{-1} contains both non-freezing water and free water does not form liquid crystal.

Figure 3-83 shows ^{23}Na longitudinal and transverse relation times (T_1 and T_2) as a function of water content (W_c) of NaCS-water systems at various temperatures. Two types of transverse relaxations are observed. One is slow and the other is fast relaxation, suggesting that the transverse relaxation decays in a non-exponential manner. In this case, the transverse relaxations produced by a quadrapole interaction are the sum of two or more decaying exponentials. As shown in Figure 3-82, the longer transverse relaxation time (T_{2s}) decreases with decreasing temperature, although the shorter relaxation time (T_{2f}) does not change much with temperature. A sudden decrease of T_{2s} values at around 260 K observed for the system with W_c higher than 1.2 g g^{-1} corresponds to the crystallization of water.

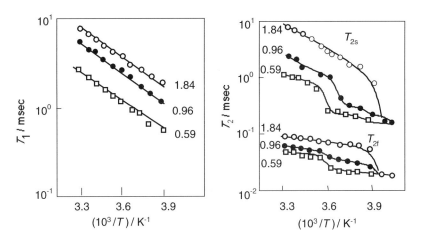

Figure 3-82. Relationships between ^{23}Na longitudinal (T_1) (left A) and transverse relaxation times and (T_2) (right B) and inverse absolute temperature of NaCS-water systems with various water contents. Numerals in the figure show water content (g g^{-1}).

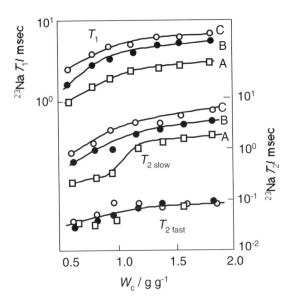

Figure 3-83. ^{23}Na longitudinal and transverse relation times (T_1 and T_2) as a function of water content (W_c) of NaCS-water systems at various temperatures. A: 293 K, B: 283 K, C: 268 K.

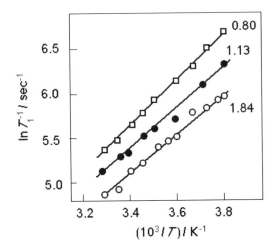

Figure 3-84. Relationships between $\ln T_1^{-1}$ and inverse absolute temperature of ^{23}Na in the NaCS-water systems. Numerals in the figure show water content (g g^{-1}).

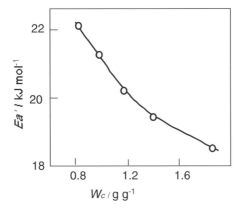

Figure 3-85. Relationship between apparent activation energy (E_a') of NaCS-water systems and water content.

From the slopes of the straight lines shown in Figure 3-84, the apparent activation energy (E_a') of the relaxation process was calculated. As shown in Figure 3-84, E_a values decrease with increasing E_a'. The result suggests that the molecular motion of ^{23}Na is strongly restricted in the low W_c region. This can be explained if it is assumed that the sodium ion is surrounded by non-freezing water and that this non-freezing water is surrounded by free water. With decreasing W_c, water molecules around the non-freezing water decrease and form hydrogen bonds between each other. This restricts the motion of the non-freezing water and indirectly that of the sodium ion. This result explains the prominent increase of E_a' of ^{23}Na in the NaCS-water system.

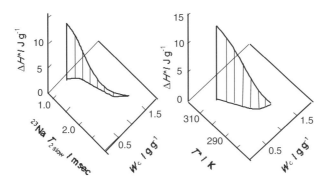

Figure 3-86. Relationship between transition temperature from liquid crystal to liquid state (T^*), transition enthalpy from liquid crystal to liquid state (ΔH^*) and water content (A) relationship between T_s slow of ^{23}Na, ΔH^* and water content (B) and of NaCS-water systems.

Figure 3-86 shows comparison between results obtained by NMR relaxation and DSC. As shown in Figure 3-86 (A) and (B), three dimensional representation among W_c, T_{2slow} at T^* and enthalpy of transition ΔH^* represents a similar pattern to the relationship among ΔH^*, T^* and W_c.

5.   HYDROGELS

A large number of polysaccharides, either polyelectrolytes or neutral ones, are known to form network structures in aqueous media. Cross-linking formation of polysaccharide hydrogels has recently received particular attention and a large number of papers have been published. Papers concerning non-cellulosic polysaccharides will be found in the references in chapter 4.

Cellulose is a water insoluble polysaccharide, accordingly, cellulose does not form hydrogels in the original form. Cellulose derivatives, such as hydroxypropylcellulose (chemical structure, see Table 3.2) and methyl-cellulose, are known to form unique hydrogels. Hydroxypropylcellulose forms lyotropic liquid crystalline phases. Various colours are visually observed when concentrated aqueous solution of hydroxypropylcellulose remains at room temperature for a certain time [90, 91]. Cholesteric liquid crystalline phase is also found in (actoxyprpoyl)cellulose [92]. Aqueous solution of methylcellulose is transparent at around room temperature, however, when the solution is heated to 343 to 353 K, it forms hydrogels. Sol-gel transition occurs reversibly depending on molecular weight and concentration.

In this section, thermal properties of natural cellulose suspension in pseudo gels, sol-gel transition of methylcelluose and chemically crosslinked carboxymethylcellulose hydrogels are described.

5.1   Cellulose pseudo gels

It is thought that water molecules play an important role in the junction zone of polysaccharide hydrogels which show unique functionality [93]. The amount of freezing bound water restrained by hydrogels which is calculated based on the melting and crystallization enthalpies of ice, is higher than that of hydrophilic linear polymers.

Cellulose is water insoluble, however, it retains a large amount of water in never dried conditions if each micro-fibre is suspended in aqueous solution. Never-dried cellulose fibres suspended in water remain in the same condition and scarcely aggregate with each other in the presence of an excess amount of water. The above cellulose suspension appears to be a

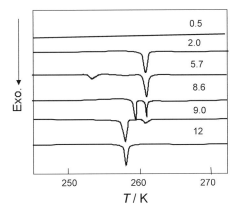

Figure 3-87. Stacked DSC cooling curves of water restrained by cellulose pseudo gels. Numerals in the figure show water content (W_c, g g^{-1}). Experimental conditions; Sample: Micro-fibril cellulose was prepared by explosion of wood pulp. The sample was supplied as a dope. The sample was dispersed in water using an ultrasonic oscillator at 298 K. The water content of the above sample was more than 100 g g^{-1}. Water in the above sample was slowly evaporated stepwise and the sample was sealed into a DSC aluminium pan. Never dried sample was used. Power compensation DSC (Perkin Elmer) was used, sample mass = 2.5 mg, cooling rate = 10 K min^{-1}. The sample was cooled from 320 K to 150 K and heated from 150 K to 320 K [94].

pseudo gel. However, when the amount of water is gradually decreased, fibres coaggregate, i.e. intermolecular hydrogen bonds are formed and water molecules surrounding each cellulose fibre are excluded. Structural change of water in the suspension during the above process can be observed by DSC.

Figure 3-87 shows stacked DSC cooling curves of water restrained by cellulose pseudo gels. Exothermic peaks are attributed to the crystallization of water in the system. Crystallization peaks vary as a function of water content in an unusual manner. In a W_c higher than ca 9 g g^{-1}, the low temperature side peak (T_{pI}) is observed. In a W_c ranging from ca. 9 to 5 g g^{-1}, two crystallization peaks, T_{pII} and T_{pI} are observed. In a W_c smaller than 5.0 g g^{-1}, thermal behaviour of water is similar to that of ordinary cellulose.

Figure 3-88 shows DSC heating curves of cellulose pseudo gels. Peaks overlap in the melting curves and a large sub peak is observed in the low temperature side of the main melting peak of water. The samples shown in this figure were heated at 10 K min^{-1} after cooling from 300 K to 200 K at the cooling rate of 10 K min^{-1}. When heating rate was decreased, the size of the low temperature side peak decreased. This fact suggests that the low temperature peak is attributable to the melting of unstable ice whose amounts are markedly influenced by thermal history.

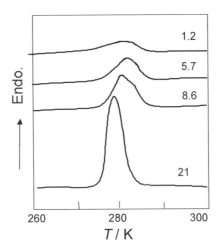

Figure 3-88. Stacked DSC heating curves of water restrained by cellulose pseudo gels. Experimental conditions (see Fig. 3-87 caption), heating rate = 10 K min^{-1}.

In order to clarify the presence of two types of freezing water in the gels, isothermal crystallization of water was carried out. Figure 3-89 shows isothermal crystallization curves of water in micro-fibre cellulose suspension with W_c = 8.9 g g^{-1}. As shown in DSC cooling curves in Figure 3-89, two crystallization peaks were observed at 258 K at 8 and 24 sec. At 261 and 262 K, one exothermic peak attributable to the short time peak is observed. The time where the peak is observed shifted to the long time side with increasing crystallization temperature. A large exothermic peak observed at longer time in isothermal crystallization curve at 258 K could not be observed at a temperature higher than 260 K. When the isothermal crystallization temperature decreases lower than 257 K, crystallization started during cooling. Hence, the temperature range capable of observing the two crystallization peaks was limited. From this figure, it is seen that the amount of water crystallized at the long time side is large.

Immediately after the completion of isothermal crystallization, each sample was heated at 10 K min^{-1}. Figure 3-89 shows DSC heating curves of the sample with W_c = 8.9 g g^{-1}. The high temperature side melting peak increases gradually with increasing crystallization temperature. At the same time, the amount of water restrained by micro fibres is far larger than bulk water which is detected as the high temperature melting peak for the sample with W_c = 0.9 g g^{-1}.

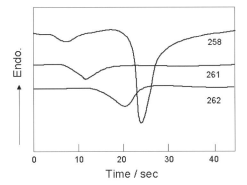

Figure 3-89. Isothermal crystallization curve of cellulose with $W_c = 8.9$ g g^{-1}. Numerals show temperature of isothermal crystallization. The samples were heated to 320 K, quenched to the temperature shown in the figure and maintained for several minutes.

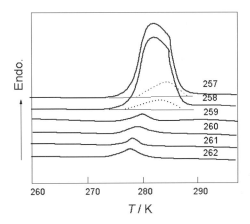

Figure 3-90. DSC melting curves of water restrained by cellulose pseudo gels immediately after completion of isothermal crystallization at various temperatures. Numerals in the figure show the temperature of isothermal crystallization. Heating rate = 10 K min^{-1}.

However, when W_c decreased to 5.6 g g^{-1}, the amount of water observed at the high temperature side increased markedly. As shown in Figure 3-91, free water crystallizes immediately after the temperature attained the isothermal state and then a small peak attributed to freezing bound water crystallizes. When crystallization curves shown in Figure 3-91 are compared with those shown in Figure 3-90, it is apparent that the amount of free water increases when water content decreases. The above facts indicate the state of water changes considerably according to the amount of water in the pseudo gels.

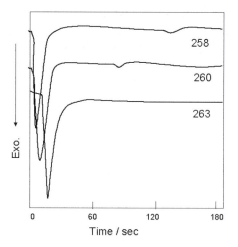

Time / sec

Figure 3-91. Isothermal crystallization curves of cellulose with W_c = 5.6 g g^{-1}. Numerals show temperature of isothermal crystallization. The samples were heated to 320 K, quenched to the temperature shown in the figure and maintained for several minutes.

   The amount of non-freezing water calculated from the melting enthalpies increased with increasing W_c, as shown in Figure 3-91. It is noted that the amounts of non-freezing water is far larger than those of natural cellulose whose W_{nf} values are 0.1 to 0.2 g g^{-1} depending on the crystallinity of the sample. From the above results, the micro fibre cellulose in the pseudo gels retains a large amount of bound water. Especially when the W_c is large, the amount of freezing bound water is larger than that of free water. In the dehydration process, cellulose molecules attach to adjacent molecules and form inter-molecular hydrogen bondings. By the formation of linkage, an

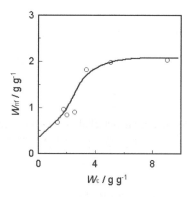

Figure 3-92. Non-freezing water content restrained by cellulose pseudo gels as a function of water content.

excess amount of water located around each cellulose molecule is excluded from the inter-molecular space and then behaves as free water. When cellulose molecules coaggregate, pseudo gels change into cellulose sheets and the number of bound water molecules approaches ordinary values.

5.2   Sol-gel transition of methylcellulose hydrogels

Among a large variety of physical hydrogel forming polysaccharides, methylcellulose (MC) shows unique characteristics [95-104], i.e. aqueous solution of methylcellulose is a transparent viscous solution at room temperature, however it forms hydrogels at a temperature higher than ca. 330 K. The gellation mechanism of MC is thought to be similar to that of the cloud point of nonionic surfactants. The cloud point increases with increasing hydrophylicity of sample. Gel-sol transition of MC increases with increasing concentration and decreasing molecular mass. Gelation of MC is thought to proceed in two stages, the first stage is hydrophobic interaction and the second stage is phase separation. By heating MC solution, water molecules attached to the hydroxyl groups of MC freely separate from the molecular chains by thermal enhancement. MC chains associate via hydrophobic interaction and three dimensional networks are gradually established [6-12]. By successive heating, phase separation occurs by gellation accompanied with water separation. The above thermally reversible gel-sol transition can be observed by visual observation using a test tube and water bath (Figure 3-93). Gel-sol transition of MC is also observed by various techniques, especially rheological and thermal measurements.

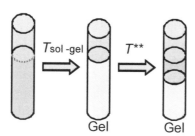

Figure 3-93. Schematic illustration of methylcellulose aqueous solution in sol, gel and phase separated state. $T_{sol-gel}$: transition temperature from sol to gel state, T^{**}: temperature of phase separation.

The transition temperatures can be determined by visual observation. When molecular weight is small, transition temperature is difficult to detect. Figure 3-94 shows DSC heating curves showing gel-sol transition (the first stage) of MC aqueous solution with various molecular weights and concentrations.

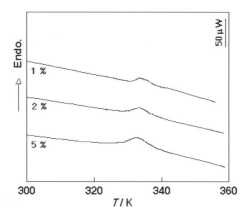

Figure 3-94. DSC heating curves of MC samples having various concentrations. Viscoity of MC samples (Wako Pure Chemicals) were 4000 cp. (Iijima, M, private communication).

5.3   Carboxymethylcelluose hydrogels crosslinked via urethane linkage

Thermal properties of carboxymethylcellulose (CMC) in dry and wet state have been described in the previous sections (2.3 and 4.3) of this chapter. Here, phase transition behaviour of chemically cross-linked CMC hydrogels is explained.   Crosslinking of CMC was carried out via the procedure as shown in Figure 3-95.   CMC-based polyurethane hydrogels were prepared in water using hexamethylenediisocyanate (HDI). Urethane linkage was formed in aqueous media using HDI.   NCOH/OH ratio was used as an index of cross-linking density. Schematic structure of CMC based polyurethane (CMCPU) hydrogels is shown in Figure 3-95. Water content ($=W_c=$ (mass of water)/(mass of dry gel), g g^{-1}) of hydrogels was varied from 0 to 10 g g^{-1}.

Phase transition behaviour of CMC hydrogels with various water contents was measured by DSC. Glass transition, melting peaks with a shoulder in the low temperature side of melting peak were observed, similar to CMC-water systems (see 3.1 of this chapter),.   Glass transition temperature of CMC hydrogels was higher than that of CMC-water systems in a water content ranging from 0 to 4.0 g g^{-1} as shown in Figure 3-96. This suggests that free molecular motion is restricted by crosslinking. Transition from liquid crystalline state to liquid state which was observed in CMC-water systems could not be found in CMCPU hydrogels. It is thought that

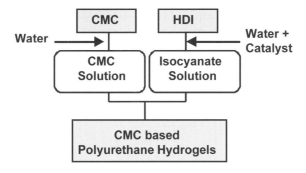

Figure 3-95. Preparation scheme of CMC based polyurethane hydrogels.

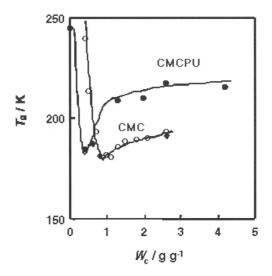

Figure 3-96. Relationship between glass transition temperature (T_g) and water content of CMC and CMC-based polyurethane hydrogels. CMC: Na carboxymethylcellulose, CMCPU: CMC-based polyurethane hydrogels, NCO/OH ratio = 1.2. Measurements; heat-flux type DSC (Seiko Instruments), heating rate = 10 K min^{-1}, sample mass = ca. 3 mg, N$_2$ flow rate = 30 ml min^{-1}.

molecular rearrangement at a temperature higher than melting of ice in the systems is suppressed by chemical crosslinking.

The effect of crosslinking is clearly seen when phase transition behaviour of CMC and that of CMC based PU is compared, as shown in Figure 3-96 The following differences are clearly observed, (1) glass transition of hydrogels is higher than that of original CMC due to crosslinking, (2) cold crystallization is not observed for hydrogels, (3) low temperature side melting peak is prominent for hydrogels, which indicates that irregular ice tends to form in the cross-linking networks (4) in contrast liquid crystal is not formed for hydrogels.

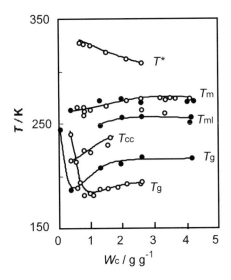

Figure 3-97. Phase diagrams of CMC and CMC-based polyurethane hydrogels. Open circles; CMC-water systems, Filled circles; CMCPU-water systems.

Dynamic modulus (E'), dynamic loss modulus (E'') and tan δ of hydrogels were measured by TMA at 298 K in water. E' of CMC hydrogels ($W_c = 6.0$ g g^{-1}) decreased with increasing NCO/OH ratio, suggesting that the amount of water restrained by CMCPU having bulky molecular chains affects the chain stiffness of the network structure. AFM images of CMC and hydrogels also support the above interpretation.

6.   THERMAL DECOMPOSITION OF CELLULOSE AND RELATED COMPOUNDS

6.1   Thermal decomposition behaviour of mono- and oligosaccharides and cellulose

Thermal degradation behaviour of mono- and oligosaccharides has been investigated by pyorysis, thermogravimetry, and simultaneous TA such as TG-mass spectrometry, TG-FTIR etc. [105-107]. Chemical structures of representative mono- and oligosaccharides, such as glucose, methyl-α-D-glucopyranoside, cellobiose and sucrose are shown in Figures.3-98. TG curves of glucose, methyl-α-D-glucopyranoside, cellobiose, sucrose and cellulose are shown in Figure 3-99. Two step decomposition was observed for mono- and oligosaccharides in N_2 atmosphere when TG derivative curves were examined. Thermal decomposition temperature (T_d) was defined in chapter 2 (Figure 2-3 of Chapter 2). Decomposition temperatures (T_d) of the first stage decomposition of the above samples and cellulose measured in N_2 at various heating rates are shown in Figure 3-100.

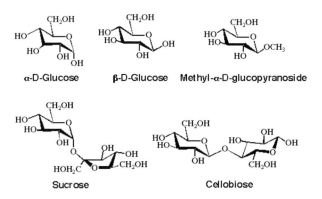

Figure 3-98. Chemical structures of glucose, cellobiose, methyl-α-D-glucopyranoside and sucrose.

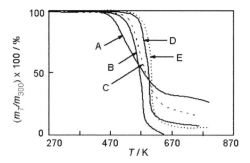

Figure 3-99. TG curves of mono- and oligosaccharides.   A: α-D-glucose, B: α-D-glucopyranoside, C: sucrose, D: cellobiose, E: cellulose. Measurements; TG (Seiko Instruments), sample mass = ca. 5 mg, heating rate = 20 K min⁻¹, N_2 flow rate = 100 ml min⁻¹.

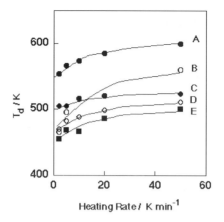

Figure 3-100. Thermal decomposition temperature (T_d) of D-glucose (E), sucrose (D), cellobiose (C) methyl-α-D-glucopyranoside (B), and cellulose (A) in N_2 atmosphere. Samples; glucose, cellobiose and sucrose were purchased from Sigma Co. Methyl- α -D-glucopyranoside was from Aldrich Chem. Co. Measurements; Thermogravimeter (Seiko Instruments TG-DTA 5200.), sample mass = ca. 5 mg, heating rate = 20 K min⁻¹, gas flow rate = 100 ml min⁻¹.

Evolved gas during thermal decomposition of various kinds of saccharides was measured by TG-DTA-FTIR. Figure 3-101 shows representative FTIR curves of D-glucose at various temperatures. The absorption bands of 670, 2350 and 3740 cm⁻¹, assigned to CO_2 and H_2O are observed. The absorption band of 670 cm⁻¹ is prominent at aroud 670 K, suggesting that the second step decomposition shown in the TG curve in Figure 3-99 is attributed to evolution of CO_2.

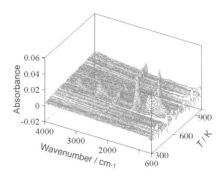

Figure 3-101. TG-FTIR curves of D-glucose measured in N_2. Samples; powder was pressed using a press machine at 100 kg cm⁻³ for 5 minutes at 298 K. A tablet having flat surfaces was placed on an open type Pt sample pan. Measurements; Thermogravimeter (Seiko Instruments TG-DTA 5200, FTIR, Japan Spectroscopic Co. Ltd., sample mass = ca. 5 mg, heating rate = 20 K min⁻¹, gas flow rate = 100 ml min⁻¹.

Microcrystalline cellulose powder is utilized in various fields, such as absorption chromatography, biocomposite fillers, non-digestible food components, matrix of pharmaceutical agents and cosmetic powder. Commercial grade cellulose powder has a wide range of size and size distribution. Microcrystalline cellulose powder samples were measured in N_2 atmosphere and air. The sample mass starts to decrease at around ca. 530 K in both in N_2 and air, however, decomposition temperature (T_d) as shown in Chapter 2, is observed at 584 K in N_2 and 569 K in air. At the same time, two step decomposition was observed for the curve measured in air. Thermal decomposition behaviour of various saccharides is found elsewhere [108].

It is known that TG curves of samples with different shapes are varied due to thermal conductivity changes in the sample pan. However, no large difference was observed among degradation temperature of cellulose powder with different sizes when powder was compressed as described in the caption of Figure 3-101. Decomposition temperature (T_d) maintains a constant value in a diameter of cellulose powder ranging from 2 to .30 μm. TG-FTIR curves of cellulose powder are shown in Figure 2-4 in Chapter 2.

6.2   Kinetics of thermal decomposition

The Ozawa-Wall-Flynn method [109, 110] for analyzing reaction kinetics using thermogravimetric (TG) data has been applied for thermal decomposition of various kinds of polymers [111]. The method is based on a simple assumption that the reaction rate [$d(x)/d(t)$, where x is reacted mass and t is time], is correlated with a function of reacted x [$f(x)$] via Arrhenius relationship. The following equation can be established between the rate of reacted mass and activation energy (ΔE).

$$dx/dt = A \exp(\Delta E/RT) f(x)$$

(3.21)

where x is mass of reaction, t time, A frequency factor, T absolute temperature and R gas constant. According to the definition of thermal analysis dT/dt = constant (=β) and β corresponds to heating rate, the following equation can be obtained,

$$\int_{x0}^{x1} \frac{dx}{f(x)} = \int_{t0}^{t1} A \exp\left(-\frac{\Delta E}{RT}\right) dt = \frac{A}{\beta} \int_{T0}^{T1} \exp\left(-\frac{\Delta E}{RT}\right) dT$$

(3.22)

when F(x) is defined as

$$F(x) = \int \frac{dx}{f(x)} \tag{3.23}$$

and partial integration is carried out one time. The equation (3.22) is transformed as

$$F(x_1) - F(x_0) = \frac{A\Delta E}{\beta R} \left\{ P\left(\frac{\Delta E}{RT_1}\right) - P\left(\frac{\Delta E}{RT_0}\right) \right\} \tag{3.24}$$

where

$$y = \frac{\Delta E}{RT}, \quad p(y) = \frac{ey}{y} - \int_y^\infty \frac{ey}{y} dy \tag{3.25}$$

As $T_1 > T_0$, the 2nd term of equation (3.24) can be neglected and the following equation can be obtained.

$$\frac{A\Delta E}{\beta R} P\left(\frac{\Delta E}{RT_1}\right) = F(x_1) - F(x_0) \tag{3.26}$$

x_1 is determined by assuming $x_0 = 0$

$$\frac{A\Delta E}{\beta R} P\left(\frac{\Delta E}{RT_1}\right) = const \tag{3.27}$$

When β is changed, only T_1 at x_1 varies. Therefore, the left hand side of equation (3.26) is constant. Using the approximation equation, equation (3.28) can be obtained.

$$\log \beta = -0.4567 \frac{\Delta E}{R} \cdot \frac{1}{T1} + const \tag{3.28}$$

If inverse T_i is plotted against log β, E can be obtained from the gradient.

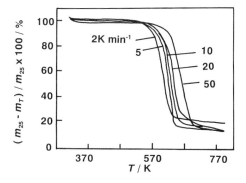

Figure 3-102. TG curves of cellulose (cotton) fabrics measured at various heating rates in N_2 atmosphere. Numerals in the figure show heating rate (K min^{-1}. Samples; JIS L 0803, cotton plane fabric No. 3,. Warp was 20 tex and weft was 16 tex. Measurement; Apparatus; (Seiko Instruments TG/DTA220, SSC 5200), Sample mass ca. 10 mg, heating rate was varied from 2 to 50 K min^{-1}, temperature range was from 303 to 790 K, Air and N_2 gas flow rate 100 ml min^{-1}.

Figure 3-102 shows TG curves of cotton fabric samples measured in N_2 at various heating rates (β). It is clearly seen that thermal decomposition occurs in one stage. The above was confirmed by the fact that a monotonous peak was observed in TG derivatograms. T_d defined in the experimental section, increases with increasing heating rate. The first 20 % mass decrease occurs similar to those in a temperature lower than 593 K. Figure 3-103 shows curves of the same samples measured in air. Thermal decomposition takes place in two stages at around 573 K and 673 K.

By TG-Fourier transform infrared spectrometry (TG-FTIR), the gas evolved during thermal decomposition was reported [112]. In a temperature lower than 573 K, two small IR absorption bands were detected at around 2300 cm^{-1}. This indicates that CO_2 is gradually evolved. On this account, in this study, the Ozawa-Wall-Flynn method was applied for the temperature range lower than 573 K. The activation energy calculated using equation (7) is shown in Table 3-10.

Table 3-8. Activation energy of initial stage of decomposition

Sample	Gas	ΔE J mol^{-1}
Cotton fabric	N_2	136
Cotton fabric	Air	125
Cellulose powder	N_2	190
Cellulose powder	Air	179

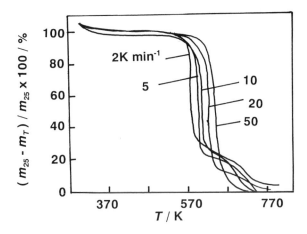

Figure 3-103. TG curves of cellulose (cotton) fabrics measured at various heating rates in air. Experimental conditions; see Figure 3-101 caption.

6.3   Life time prediction of cellulose

Aging is an important property when green polymers are practically used [113]. Using the above kinetic data, the life-time of polymeric materials can be estimated [114]. When the temperature is assumed to be held at a constant, T_c, the equation (3-21) can be rewritten as follows.

$$\frac{dx}{dt} = A \exp\left(-\frac{\Delta E}{RT_c}\right) f(x) \tag{3.29}$$

As T_c is independent of time (t) and life-time τ is assumed to be defined as the time interval in which x changes from x_0 to x_1, the equation (9) can be obtained.

$$\int_{x_0}^{x_1} \frac{dx}{f(x)} = A \int_{t_0}^{t_0+\tau} \exp\left(-\frac{\Delta E}{RTc}\right) dt = A \exp\left(-\frac{\Delta E}{RT_c}\right) \tau \tag{3.30}$$

By comparing the above equation to equation (3-22), the life-time τ is obtained as follows.

$$\tau = \frac{\int_{T\,0}^{T\,1}\exp\left(-\dfrac{\Delta E}{RT}\right)dT}{\beta\left(-\dfrac{\Delta E}{RTc}\right)} \qquad (3.31)$$

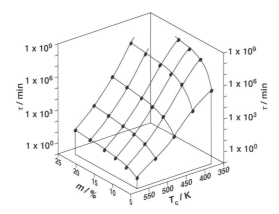

Figure 3-104. Three dimensional relationship between life-time, τ. (minutes), holding time, T_c, (absolute temperature) and mass decrease, m, (%) of cotton fabric in N_2 [114].

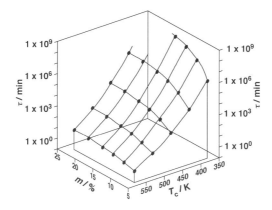

Figure 3-105. Three dimensional relationship between life-time τ (minutes), holding time T_c, (absolute temperature) and mass decrease m, (%) of cotton fabric in air [114].

Figures 3-104 and 3-105 show the three dimensional relationship between τ, T_c and mass decrease of cotton fabric in N_2 and air. It is clearly seen that τ varies from 1 sec to 100 years depending on T_c. At the same time, decomposition is delayed by maintaining the fabric in an inert atmosphere, such as N_2. Cellulose powder shows almost similar relationships as shown in Figures 5- 104 and 5-105. The difference between N_2 and air increases with decreasing T_c. As shown in Figure 6, 10 % mass decrease of cotton fabrics at 373 K takes 72 years in air and 150 years in N_2.

REFERENCES

1. Colvin, J. R. 1985,Cellulose biosynthesis, in *Encyclopedia of Polymer Science and Engineering* vol 3, 60-68, John Wiley & Sons, New York.
2. Daniel, J. R., 1985, Cellulose, structure and properties, in *Encyclopedia of Polymer Science and Engineering* vol 3, 90-123, John Wiley & Sons, New York.
3. Soc. Cellulose Chemistry ed., 2000, *Cellulose Handbook*, Asakura Pub., Tokyo (Japanese).
4. Hermans, P. and H., Weidinger, A., 1949, X-ray studies on the crystallinity of cellulose, *J. Polym. Sci.*, **4**, 135-144.
5. Hermans, P. H. and Weidinger, A., 1949, Estimation of crystallinity of some polymer from x-ray intensity measurements, *J. Polym. Sci.*, **4**, 709-723.
6. Marchesault, R. H. and Howsmon, J. A., 1957, Experimental evaluation of the lateral-order distribution in cellulose, *Text. Res. J.*, **27**, 30-41.
7. Komatsu, N. and Sakata, A., 1958, Preparation of amorphous cellulose and crystallinity measurement, *Kogyo Kagaku Zasshi, (Industrial Chemistry J.)*. (Japanese).
8. Mann, J., Roldan-Gonzalez, L. and Weillard, H. J., J., 1960, Crystalline modification of cellulose IV Determination of x-ray intensity data, *J. Polym. Sci.*, **42**, 165-171.
9. Ellis, K. C. and Warwicker, J. O. 1962, A study of the crystal structure of cellulose I, *J. Polym. Sci.*, **56**, 339-357.
10. Chang, M., 1971, Holding chain model and annealing of cellulose, *J. Polym. Sc., Part C*, **36**, 343-352.
11. Gardener, K. H. and Blackwell, J., 1974, The structure of native cellulose, *Biopolymers*, **13**, 1975-2001.
12. Kulshereshtha, A. K., Chudasama, U. P. and Dweltz, V. E., 1975, Analysis of cotton fiber maturity. I. X-ray study of phase transformation in various cottons, *J. Appl Polym. Sci.*, **19**, 115-123.
13. Kolpak, F. J., and Blackwell, J., 1976, Determination of the structure of cellulolse II, *Macromolecules*, **9**, 273-278.
14. Sarko, A., 1976, Crystalline polymorphs of cellulose: Prediction of structures and properties, *Appl. Polym. Sym.*, **28**, 729-743
15. Pertsin, A. T., Nugmanov, O. K., Sopin, V. F., Martchenko, G. N. and Kitaigorodskii, A. I., 1981, Conformation of isolated cellulose helix, *Visokomol. Soeg.*, **23**, 2147-2155 (Russian).
16. Zugenmaier, P., structure investigations of cellulose derivatives, 1983, *J. Appl. Polym. Sci.*, **37**, 223-238.
17. Horii, F., Hirai., A. and Kitamaru, R., 1984, Cross-polarization/magic angle spinning ^{13}C-NMR study, molecular chain of native and regenerated cellulose, in *Polymers of Fibers and Elastomers*, Arthur J. C., ed., ACS Symp. Ser., 260, A. Chem., Soc., Wahington D.C., 27-42.
18. Kataoka, Y. and Kondo, T., 1998, FT-IR microscopic analysis of changing cellulose crystalline sturcutre during wood cell wall formation., *Macromolecules*, **31**, 760-764.

19. Hainze, T. J. and Glasser W. G., 1996, Cellulose derivatives, modification, characterization and nanosturctures, *ACS Symp. Ser.,* **688**, Am. Chem. Soc., Washington DC.

20. Mikhailow, G. P., Artukhov, A. I. and Shevelv, V. A., 1969, Study on molecular motion of cellulose and its derivatives by dielectric and NMR methods, *Vysokomol. Soeg.,* **11**, 553-563 (Russian).

21. Hatakeyama, T. Nakamura, K. and Hatakeyama, H., 1982, Studies on heat capacity of cellulose and lignin by differential scanning calorimetry, *Polymer,* **23** 1801-1804.

22. Kamide, K. and Saito, M., 1985 Thermal analysis of cellulose acetate solids with total degrees of substitution of 0.49, 1.75, 2.46 and 2.92. *Polymer J.,* **17**, 919-928.

23. Kamide, K., Okajima, K., Kowsaka, K., Matsui, T., Nomura, S. and Hikichi, K., 1985, Effect of the distribution of substitution of the sodium salt of carboxymethylcellulose on its absorbency toward aqueous liquid, *Polym. J.,* **17**, 909-918.

24. Nakamura, K., Hatakeyama, T. and Hatakeyama H., 1996. Heat capacity of the carboxymethylcelluose-water system near the glass transition", Kobunshi Ronbunsyu (*J. Soc. Polym. Sci. Technol.*) **53** (12), 860-865 (Japanese).

25. Hatakeyama, H. and Hatakeyama, T., 1974, Formation of hydrogen bonding by heat-treatment of amorphous cellulose. *Sen-i Gakkaishi, (J. Soc. Textile Sci. Technol. Japan),* **30**, T214-T220.

26. Hatakeyama, H. Hatakeyama, T. and Nakano, J., 1974, Studies on hydrogen-bond formation of amorphous cellulose, *Cellulose Chem. Technol.,* **8**, 495-509.

27. Yano, S., Hatakeyama, H. and Hatakeyama, T., 1976, Effect of hydrogen bond formation on dynamic mechanical properties of amorphous cellulose. *J. Appl. Polym. Sci.,* **20**, 3221-3231.

28. Hatakeyama, H., Hatakeyama, T. and Nakano, J., 1976, The rffect of hydrogen bond formation on the structure of amorphous cellulose, *J. Appl. Polym. Sci., Symp.,* **28**, 743-750.

29. Hatakeyama, H. and Hatakeyama, T., 1981, Structural change of amorphous cellulose by water- and heat-treatment, *Makromol. Chem.,* **182**, 1655-1668.

30. Alfthan, E. and deRuvo A., 1973, Glass transition temperatures of oligosaccharides, *Polymer* **14**, 329-330.

31. Hatakeyama, H., Yoshida, H., Nakano, J., 1976, Studies on the isothermal crystallization of D-glucose and cellulose oligosaccharides by differential scanning calorimetry", *Carbohydrate Res.,* **47**, 203-211.

32. Hatakeyama, T., Yoshida, H., Nagasaki, C. and Hatakeyama, H., 1976, Differential scanning calorimetric studies on phase transition of glucose and cellulose oligosaccharides", *Polymer,* **17**, 559-562.

33. Hatakeyama, H. and Hatakeyama, T., 1977, Fractionation of cellulose oligosaccharides by gel permeation chromatography", *Mokuzai Gakkaishi,* **23**, 228-231.

34. Hatakeyama, H., Nagasaki, C. and Yurugi, T., 1976, Relation of certain infrared bands to conformational changes of cellulose and cellulose oligosaccharides", *Carbohydrate Research,* **48**, 149-158.

35. Franks, F ed., *Water, A Comprehensive Ttreatise,* 1972, vol. 1 The physics and physical chemistry of water, vol 2, Water in crystalline hydrates, Aqueous solutions of simple nonelectrolytes, vol. 3 Aqueous solutions of simple electrolytes, vol.4 Aqueous solutions of amphiphiles and macromolecules, vol. 5 Water in disperse systems, vol 6, Recent advances, vol 7, Water and aqueous solutions at subzero temperatures, Plenum press, New York.

36. Errede, L. A., 1991, Molecular interpretations of sorption in polymers, Part I, Springer-verlag, Berlin.
37. Waigh, T. A., Jenkins, P. J. V. and Donald, A. M., 1996, Quantification of water in carbohydrate lamellae using SANS, *Faraday Discuss.*, **103**, 325-337.
38. Hatakeyama, H. and Hatakeyama, T., 1998, Water-Polymer Interaction, *Thermochimica Acta* **308**, 3-22.
39. Hatakeyama, T., Ikeda, Y. and Hatakeyama, H., 1987, Structural change of the amorphous region of cellulose in the presence of water", "Wood and Cellulose", (Eds. J. F. Kennedy et al.), Ellis Horwood Ltd., Chichester, Ch.2, 23.
40. Magne, F. C., Fortas, H. J. and, Wakeham, H., 1947, A calorimetric investigation of moisture in textile fibers, *J. Am., Chem., Soc.*, **69**, 1896-1902.
41. Preston, J. M. and Tawde, G. P., 1956, Freezing point depression in assemblages of moist fibres, *J. Tex. Inst*, **47**, T157-65.
42. Neal, J. L. and Goring, D. A. I., 1969, Interaction of cellulose with liquid water accessibility, determination from thermal expansion, *J. Polym. Sci., Part C*, **28**, 103-113.
43. Newns, A. C., 1973, Sorption and desorption kinetics of the cellulose and water system. Part2. Low concentration on the lowest sorption limits of the hysteresis loop, *J. Chem. Faraday I*, **64**, 444-448.
44. Froix, M. F. and Nelson, R., 1975, The interaction of water with cellulose from nuclear magnetic resonance relaxation times, *Macromolecules*, **8**,726-730.
45. Nelson, R. A., 1977, The determination of moisture transition in cellulosic materials by differential scanning calorimetry, *J. Appl. Polym. Sci.*, **21**, 645-654.
46. Hatakeyama, T., Nakamura, K. and Hatakeyama, H., 1979, Determination of bound water contents absorbed on polymers by differential scanning calorimetry, *Netsu*, **6**, 50-52. (Japanese)
47. Hatakeyama, T., Ikeda, Y. and Hatakeyama, H., 1987, Effect of bound water on structural change of regenerated cellulose. *Makromol. Chem.*, **188**, 1875-1884.
48. Nakamura, K., Hatakeyama, T. and Hatakeyama, H., 1981, Studies on Bound Water of Cellulose by Differential Scanning Calorimetry, *Textile Res. J.*, **51**, 607.
49. Nakamura, K., Hatakeyama, T. and Hatakeyama, H., 1983, Effect of Bound Water on Tensile Properties of Native Cellulose, *Textile Res. J.*, **53**, 682-688.
50. Nakamura, K., Hatakeyama T. and Hatakeyama, H., 1981, Relationship Between Official Regain and Bound Water in Cellulose, *Sen-i Gakkaishi* (*J. Soc. Fibre Scoi, and Technol. Japan*), **37**, T533-T535.
51. Hatakeyama, T., Nakamura, K. and Hatakeyama, H., 1990, Thermal analysis of bound water in polysaccharides, in *Cellulose Sources and Exploitation* (J. F. Kennedy et al. Eds.), Ellis Horwood, Chichster, Ch 2, 13-19.
52. Hatakeyama, T. and Hatakeyama, H., 1985, Heat capacity of water-cellulose and water-oligosaccharides systems, in *Cellulose and its Derivatives: Chemistry, Biochemistry and Applications* (J. F. Kennedy et al. Eds.), Ellis Horwood, Chichester, Chapter, 7, 87-94.
53. Pyda, M., 2001, Conformational contribution to the heat capacity of the starch and water system, *J. Polym. Sci., Polym. Phys.*, **39**, 3038-3054.
54. Pyda, M., 2002, Conformational heat capacity of interaction systems of polymer and water, *Macromolecules*, **35**, 4009-4016.
55. Hatakeyama, T., Nakamura, K. and Hatakeyama, H., 1988, Determination of bound water content in polymers by DAT, DSC and TG, *Thermochimica Acta*, **123**, 153-161.
56. Hatakeyama, T., Nakamura, K. and Hatakeyama, H., 2000, Vaporization of bound water associated with cellulose fibres, *Thermochimica Acta*, **352-353**, 233-239.

57. Yano, S. and Kitano, T., 1996, Dynamic viscoelastic properties of polymeric materials, in *Handbook of Applied Polymer Processing Technology*, Chapter 4, Cheremisinoff N. P., Cheremisinoff P. N., eds. Marcel Dekker Inc.

58. Yano, S., Hatakeyama, H. and Hatakeyama, T., 1976, Effect of hydrogen bond formation on dynamic mechanical properties of amorphous cellulose, *J. Appl. Polym. Sci.*, **20**, 3221-3231.

59. Manave, S., Iwata, M. and Kamide, K., 1986, Dynamic mechanical absorptions observed for regenerated cellulose solids in the temperature range from 20 to 600 K, *Polym. J.*, **18**, 1-14.

60. Yano, S. and Hatakeyama, H., 1988, Dynamic viscoelasticity and structural changes of regenerated cellulose during water sorption, *Polym.* **29**, 566-570.

61. Yano, S., Hatakeyama, H. and Hatakeyama, T., 1989, The Dynamic viscoelasticity of cellulose in water, in *Cellulose and Wood* (ed. C. Schuerch) John Wiley and Sons, N. Y 389-402.

62. Zhou, S., Tashiro, K., Hongo, T., Shirataki, H., Yamane, C. and Ii, T., 2001, Influence of water on structure and mechanical properties of regenerated cellulose studied by an organized combination of infrared spectra, x-ray diffraction, and dynamic viscoelastic data measured as functions of temperature and humidity, *Macromolecules*, **34**, 1274-1280.

63. Morra, M., ed., 2001, *Water in Biomaterials Surface Sscience and Bbiomembrane*, John Wiley and Sons., Chichester.

64. Tsurumi, T., Osawa, N., Hitaka, H., Hirasaki, T., Yamaguchi, K., Manabe, S. and Yamashiki, T., 1990, Structure of cuprammonium regenerated cellulose hollow fiber (BMM Hollow fiber) for virus removel, *Polym. J.*, **22**, 751-758.

65. Boyle, N. G., McBrierty, V. J. and Douglass, D. C., A study of the behavior of water in Nafion memebranes, *Macromolecules*, **98**, 75-80.

66. Hatakeyama, T., Yamamoto, S., Hirose, S. and Hatakeyama, H., 1989, Structural change of water restrained in cellulosic hollow fibers, in *Cellulose and Wood* (ed. C. Schuerch), John Wiley and Sons, New York, 431-446.

67. Kimura, M., Hatakeyama, T. and Nakano, J., 1974, DSC study on recrystallization of amorphouys cellulose with water, *J. Appl. Polym. Sci.*, **18**, 3069-3076.

68. Hatakeyama, H., Hatakeyama, T. and Nakamura, K., 1983, Relationship between Hydrogen Bonding and Water in Cellulose, *Journal of Applied Polymer Science. Symposia.*, **37**, 979-991.

69. Hatakeyama, T. and Hatakeyama, H., 1992, Molecular relaxation of cellulosic polyelectrolytes with water, in *Viscoelasticity of Biomaterials* (W. G. Glasser and H. Hatakeyama eds.), ACS Symposium Series 489, ACS, Washington DC, Ch.22, 329-340.

70. Hatakeyama, T., Bahar, N. and Hatakeyama, H., 1991, Liquid Crystalline State of Water-Carboxymethylcellulose Systems Substituted with Mono- and Divalent Cations", *Sen-i Gakkaishi (J. Soc. Textile Sci. Technol. Japan)* **47**, 417-420.

71. Tsuchida, E. and Abe, K., 1986, Polyelelctrolyte complexes, in *Developments in Ionic Polymers* -2, Wilson, A. D. Prosser, H. J. eds. Elsevier Applied Science Pub., London, Chapter 5, 191-265.

72. Nakamura, K., Hatakeyama, T. and Hatakeyama, H., 1987, DSC Studies on monovalent and divalent cation salts of carboxymethylcellulose in highly concentrated aqueous solutions", in *Wood and Cellulose*" (Eds. J. F. Kennedy et al.) Ellis Horwood Ltd., Chichester, Ch. 10, 97-103.

73. Kamide, K., Yasuda, K. and Okajima, K., ^{13}CNMR study on gelation of aqueous carboxymethylcellulose with total degree of substitution of 0.39 solution induced by metal cations. *Polym. J.,* **20**, 259-268 (1988).

74. Berthold, J., Desbrières, J., Rinaudo, M. and Salmén, L., 1994, Types of adsorbed water in relation to the ionic groups and their counter-ions for some cellulose derivatives, *Polym.*, **35**, 5729-5736.

75. Bhaskar, G., Ford, J. L. and Hollinger D. A., 1998, Thermal analysis of the water uptake by hydrocolloids, *Thermochimica Acta*, **322**, 153-165.

76. Matsumoto, T. and Ito, D., 1990. Viscoelastic and nuclear magnetic resonance studies on molecular mobility of carboxymethylcellulose-calcium complex in concentrated aqueous systems. *J. Chem. Soc. Faraday Trans,* **86**, 829-832.

77. Hatakeyama, T., Hatakeyama, H. and Nakamura, K., 1995, Non-freezing water content of mono- and divalent cation salts of polyelectrolyte-water systems studied by DSC", *Thermochimica Acta*, **253**, 137-148.

78. Hoffman, K. and Hatakeyama, H., 1994, [1]H n.m.r. Relaxation studies and lineshape analysis of aqueous aodium carboxymethylcellulose, *Polymer*, **35**, 2749-2758

79. Nakamura, K., Hatakeyama, T. and Hatakeyama, H., 1996 Heat capacity of the carboxy-methylcelluose-water system near the glass transition", *Kobunshi Ronbunsyu*, **53**, 860-865 (Japanese).

80. Kamide, K. and Okajima, K., 1981, Determination of distribution of sodium sulfate groups in glucopyranose units of sodium cellulose sulfate by ^{13}C and ^{1}H nuclear magnetic resonance analysis, *Polym. J.*, **13**, 163-166.

81. Hatakeyama, H. and Hatakeyama, T., 1991, Mesomorphic properties of polyelectrolytes with water, in *Properties of Ionic Polymers Natural and Synthetic*, L. Salmen and Myat Htun (editors), STFI-meddelande, **a 989**, p123-147.

82. Hatakeyama, T., Yoshida, H. and Hatakeyama, H., 1987, A differential scanning calorimetry study of the phase transition of the water-sodium cellulose sulfate system", *Polym.*, **28**, 1282-1286.

83. Hatakeyama, H., Iwata, H. and Hatakeyama, T., 1987, ^{1}H and ^{23}Na NMR studies of the interaction between water and sodium cellulose sulfate, in *Wood and Cellulose*, (Eds. J. F. Kennedy et al.), Ellis Horwood Ltd., Chichester, Ch, 4, 39-46.

84. Hatakeyama, T., Yoshida, H. and Hatakeyama, H., 1995, The liquid crystalline state of water-sodium cellulose sulphate systems studied by DSC and WAXS. *Thermochimica Acta*, **266**, 343-354.

85. Hatakeyama, H. and Hatakeyama, T., 1990, Nuclear magnetic relaxation studies of water-cellulose and water-sodium cellulose sulfate systems, *in Cellulose: Structural and Functional Aspects* (J. F. Kennedy et al. Eds.), Ellis Horwood, Chichster, Ch 14, 131-136.

86. Hatakeyama, H., Yoshida and H. and Hatakeyama, T., 1985, Study of the Interaction between water and cellulose sulfate sodium salt by DSC and NMR, *Cellulose and it's Derivatives* (J. F. Kennedy et al. Eds.), Ellis Horwood, Chichester, Chapter, 21, 255-262.

87. Hatakeyama, H., Nakamura, K. and Hatakeyama, T., 1989, DSC and NMR studies on the water-cellulosic polyelectroyte systems, in *Cellulose and Wood* (ed. C. Schuerch), John Wiley and Sons, N.Y., 419-429.

88. Hatakeyama H. and Hatakeyama, T., 1990, Nuclear magnetic relaxation studies of water-cellulose and water-sodium cellulose sulfate system", in *Cellulose: Structural and Functional Aspects* (J. F. Kennedy et al. Eds.), Ellis Horwood, Chichester, Ch 14, 131.

89. Woessner, D. E. and Snowden, Jr., B. S., 1970, Pulsed Nmr study of water in agar gels, *J. Colloid Interface Sci.*, **34**, 290-299.

90. Werbopwyj, R. S. and Gray, D. C., 1976, Liquid crystalline structure in aqueous hjydroxypropyl cellulose solution, *Mol. Cryst. Liq. Cryst.* **34**, 97-103.

91. Werbopwyj, R. S. and Gray, D. C., 1980, Ordered phase formation in concentrated hydroxypropylcellulose solutions, *Macromolelcules*, **13**, 69-78.

92. Tseng, S-L., Valente, A. and Gray, D. C., 1981, Cholecteric liquid crystalline phase based on (acetoxylpropyl)cellulose, *Macromolecules,* **14**, 715-719.

93. Hatakeyama, T. Yoshida, H. and Hatakeyama, H., 1990 DSC study on water in polysaccharide gels, in Cellulose, Structural and Functional Aspects, (J. F.Kennedy et al. Eds.), Ellis Horwood, Chichster, Ch 39, 305-310.

94. Hatakeyama, T. and Hatakeyama, H., 1993, Thermal properties of water around the cross-linking networks in cellulose pseudo hydrogels"; in "Cellulosics: Chemical, Biochemical and Material Aspects, (J. F. Kennedy, G. O. Phillips and P. A. Williams eds.), Ellis Horwood, Chichester, Chap. 32, 225-230.

95. Kato, T., Yokoyama, M. and Takahashi, A., 1978, Melting temperatures of thermally reversible gels, IV. Methyl cellulose-water gels., *Colloid & Polym. Sci.,* **256**, 15-21.

96. Sarkar, N., 1979, Thermal gelation properties of methyl and hydroxypropyl methylcellulose, *J. Appl. Polym. Sci.,* **24**, 1073-1087.

97. Hirrien, M., Desbrieres, J. and Rinaudo, M., 1996, Physical properties of methylcelluloses in relation with the condition for cellulose modification, *Carbohydrate Polym.*, **31**, 243-252.

98. Nishinari, K. Hofmann, K. E., Moritaka, H., Kohyama, K. and Nishinari, N., 1997, Gel-sol transition of methylcellulose, *Macromol., Chem., Phys.*, **198**, 1217-1226.

99. Desbrieres, J., Hirrien, M. and Rinaudo, M., 1998, A calorimetric study of methylcellulose gelation, *Carbohydrate Polym.*, **37**, 145-152.

100. Desbrieres, J., Hirrien, M. and Rinaudo, M., 1998, Relation between the conditions of modification and the properties of cellulose derivatives* thermogelation of methylcellulose, in *Cellulose Derivatives, Mmodification, Characterization and Nanostructures,* Heinze, T J., Glasser, W. G., eds., ACS Symp. Ser., **688**, Am. Chem., Soc., Washington D. C., 332-348.

101. Ford, J. L., 1999, Thermal analysis of hydroxypropylmethylcellulose and methylcellulose powders, gels and matrix tablets. *Inter. J. Pharmaceutics*, **10**. 209-228.

102. Li., L., Thangamathesvaran P. M., Yue, C. Y., Tam, K. C., Hu X. and Lam., Y. C., 2001, Gel network structure of methylcellulose in water, *Macromolecules*, **17**, 8062-8068.

103. Rinaudo M. and Desbrieres, J., 2000, Thermally induced gels obtained with some amphiphilic polysaccharide derivatives: synthesis, mechanism and properties, in *Hydrocolloids-Part 1,* Nishinari, K. ed., Elsevier Sci., B. V. 111-123.

104. Li, L., Thermal gelation of methylcellulose in water, 2002 Scaling and thermoreversibility, *Macromolecules*, **35**, 5990-5998.

105. Shafizadeh, F., 1971, Thermal behavior of carbohydrates, *J. Polymer Sci., Part C*, **36**, 21-51.

106. Shafizadeh F. and Fu, Y. L. 1973, Pyrolysis of cellulose, *Carbohydrte Res.*, **29**, 113-122.

107. Shafizadeh F., McGinnins, G.D., Susott, R. A. Philpot C. W., 1970, Thermodynamic properties of 1,6-anahydrohexopyranose crystals, *Carbohyd. Res.*, **15**, 165-178.

108. Nguyenm T., Zavarin E. and Barrall II, E. M., 1981, Thermal analysis of lignocelllose materials, Part I. Unmodified materials, *J. Macromol. Sci. Rev.* **C20**, 1-65 (1981) Part II modified materials, C21 1-60.

109. Ozawa, T., 1965, A new method of analyzing thermogravimetric data, *Bull. Chem. Soc. Japan*, **38**, 1881-1886.

110. Flynn, J. H. and Wall, L. A., 1966, A quick, direct method for the determination of activation enery from thermograimetric data, *Polymer Letter*, **4**, 323-328.

111. Kaloustian, J., Pauli, A. M. and Pastor, J., 2001, Kinetic study of the thermal decompositions of biopolymers extracted from various plants, *J. Thermal. Anal. Cal.*, **63**, 7-20.

112. Hatakeyama T. and Liu, Z. ed., 1998, *Handbook of Thermal Analysis*, John Wiley & Sons, pp148.

113. Kleinert T. N. 1972, Aging of cellulose Pt. VI. Natural ageing of linen over long periods of time, *Holzforshung*, **26**, 46-51.

114. Hatakeyama T., Nakazawa, J., Iijima, M. and Hatakeyama, H., 2002, Thermogravimetric study on life-time prediction of cellulose fabrics, *Sen-i Gakaishi, (J. Soc. Fibre Sci. Technol. Japan)*, **58**, 405-408.

Chapter 4

POLYSACCHARIDES FROM PLANTS

Recently, functional properties of polysaccharides have received particular attention, due to the possibility of application in a wide range of fields [1-9]. Among a large number of terraqueous plants, cellulose and lignin are dominant in nature. In this book, Chapters 3 and 5 are devoted to the two major components of terraqueous plants. In this chapter, thermal properties of a selected number of carbohydrate polymers, obtained from plants, either terraqeuous or aquatic, and bacteria, will be described. Formation of hydrogels and liquid crystals of polysaccharides will be focused on from the view point of structural change of water via polysaccharide-water interaction. In order to consider practical applications, hydrogels obtained by chemical modification will also be described.

1.   GELATION

Various kinds of polysaccharides, such as agarose [10,11], carrageennan [12 - 15] gellan gum [16 - 20], konjac glucomannan [21 22], etc are known to form physical gels in aqueous media, and a there are large number of papers devoted to the gelation mechanism of polysaccharides [23 - 25]. In this section, thermal properties of representative polysaccharide hydrogels are described.

1.1   Gelation of alginic acid

1.1.1   Gel formation of alginic acid via divalent cations

Alginic acid is a copolysaccharide consisting of manuronic and guluronic acid which is extracted from seaweed, *brown masse algae* [26, 27, 28]. Chemical structures of both components are shown in Figure 4-1. Sodium alginate (NaAlg) is water soluble and has been used in a wide range of industrial fields. It is known that NaAlg forms water insoluble gels when Na ion of aqueous solution of NaAlg is replaced by Ca ions in an aqueous media, such as $CaCl_2$ aqueous solution [29]. It is thought that guluronic components enclosing Ca ions coaggregate and act as junction zones. On this account, the size of junction zones depend on manuronic/guluronic (M/G) ratio of the sample. When NaAlg aqueous solution is introduced in $CaCl_2$ solution as a fibre form, CaAlg gel fibres are formed [30]. In this case, the molecular axis aligns along the b axis and CaAlg containing a large amount of water shows liquid crystalline structure [31]. When aqueous NaAlg solution is introduced dropwise in water, spherical CaAlg gels are formed. This property is utilized to produce man-made fish eggs which are widely used in food industries

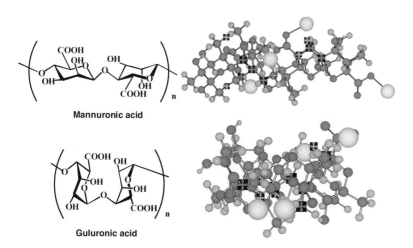

Mannuronic acid

Guluronic acid

Figure 4-1. Chemical structure of mannuronic and guluronic acid.

It is thought that physical properties of CaAlg hydrogels depend not only on M/G ratio and molecular weight, but also on gel preparation conditions, such as, concentration of NaAlg solution, Ca concentration, substitution time, and other factors. When sphere type gel is formed, Na ions are instantaneously substituted with Ca ions. After the junction zone is completed on the surface, Ca ions diffuse from the skin into the core, and this process depends on time. At the same time, it is thought that the structure of NaAlg solution affects the mobility of Ca ions.

In our recent studies on hydrogel formation of polysaccharides, it was found that equilibration of aqueous solution is an important factor. In this section, gelation of alginan and also structural change of water in alginan are described.

1.1.2   Phase transition of NaAlg-water systems

Similar to other polysaccharide electrolytes, NaAlg-water systems show unique phase transition behaviour [32, 33]. Figures 4-2.A and 4-2.B show DSC cooling and heating curves of NaAlg-water systems with various water contents. As no transition was observed for completely dry NaAlg in a temperature range from 120 K to 370 K, transitions shown in Figures 4-2.A

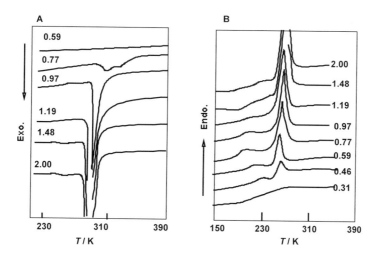

Figure 4-2. DSC curves of water-NaAlg systems. A; cooling curves, B; heating curves. Measurements; Heat flux type DSC (Seiko Instruments), scanning rate = 10 K min^{-1}, sealed type aluminium pan, sample mass = ca. 3 mg.

and 4-2.B are caused by the molecular motion attributed to water or NaAlg molecules enhanced by water. As shown in Figure 4-2.A, the system with water content 0.59 g g^{-1} shows no transition, however, in Figure 4-2.B, an endothermic peak can clearly be seen. By the comparison of cooling and heating curves, it was found that the difference was significant when water content is lower than 1.0 g g^{-1}. The difference indicates that systems solidify in glassy state and crystallization takes place during the heating process. In heating curves, a baseline gap due to glass transition and a shallow exothermic peak attributable to cold crystallization are observed.

Using transition temperatures obtained by DSC heating curves, relationship between phase transition temperature of water-NaAlg systems and W_c was established as shown in Figure 4-3. The T_g decreased in a W_c smaller than 0.5 g g^{-1}, reached a minimum at W_c = ca. 7.0 g g^{-1} and then slightly increased with increasing W_c. T_{cc} is found in a narrow W_c range. When T_m levels off, T_g also maintains a constant value.

Non-freezing water content of NaAlg was calculated using the enthalpy of melting and crystallization. As described in 3.4, the amount of free water was calculated from the following equation,

$$W_{nfc} = W_c - W_{fc} \tag{4.1}$$

$$W_{nfm} = W_c - W_{fm} \tag{4.2}$$

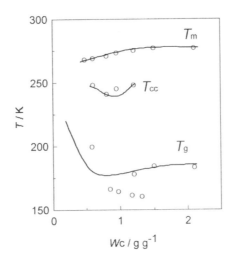

Figure 4-3. Phase diagram of water-NaAlg systems.

where W_{fc} is the amount of free water estimated by crystallization enthalpy, W_{fm} is that of melting enthalpy and two different amounts of non-freezing water, W_{nfc} and W_{nfm} were calculated. From W_{nfc} and W_{nfm}, the number of bound water molecules attached to repeating units of NaAlg was calculated as shown in Figure 4-4. It is seen that the number of water molecules estimated from the crystallization enthalpy is larger than those calculated from the melting enthalpy. This suggests that a part of non-freezing water crystallizes into irregular ice during heating. The cold crystallization observed at a temperature between T_g and T_m shown in Figures 4-2.B is attributable to the difference (W_{nfc}-W_{nfm}) found in Figure 4-4. Besides the difference of the two curves, another characteristic feature of Figure 4-4 is that the maximum point is observed at around 0.8 to 1.0 g g^{-1}. As shown in Figure 4-3, the melting peak temperature of water levels off at around W_c = 1.0 g g^{-1}. This suggests that the number of free water molecules is insufficient to form regular ice at a W_c smaller than 1.0 g g^{-1}. The fact that the number of nonfreezing water molecules reaches the maximum indicates that the amount of amorphous ice involved in molecular motion of NaAlg is the maximum. The value of heat-capacity difference at T_g (ΔC_p) is a criteria to estimate molecular motion of a sample at T_g. Figure 4-5 shows the relationship between ΔC_p and W_c. As clearly seen, the maximum is found at around W_c = 1.0 g g^{-1}. After exceeding the maximum point, ΔC_p decreases rapidly. This can be explained by the fact that the molecular motion is

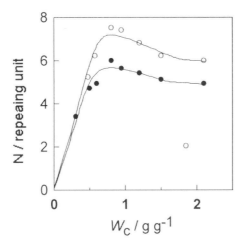

Figure 4-4. Number of water molecules restrained as non-freezing water calculated from crystallization and melting enthalpy.   Open circles; calculated from enthalpy of crystallization, closed circles; enthalpy of melting.

restricted by the presence of ice in the system at a T_g range. The ratio of non-freezing water among total water molecules decreases and molecular enhancement of NaAlg hardly occurs.

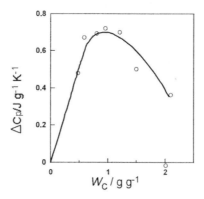

Figure 4-5. Heat capacity gap at T_g of NaAlg-water systems as a function of water content.

1.1.3   Bound water of Alginan with various counter ions

When Na ions of NaAlg are converted into divalent or trivalent cations, crosslinked water insoluble materials are obtained. Degree of substitution (DS) can be controlled by experimental conditions, such as concentration of the NaALg aqueous solution (dope), conversion time, etc. When dope having 2 - 3 % concentration was extended on a glass plate, dried at room temperature and immersed in metal chloride aqueous solutions (concentration 1-4 %), films having various DS were prepared by controlling the time for immersion. Conversion rate depends on ion species, for example in the case of $FeCl_2$, DS was maintained at around 0.1 until 10 min, then quickly increased with increasing time, and films with DS =1.0 were obtained at around 120 min. The films thus prepared show various colours depending on ion species. Thermal and mechanical properties of dry and wet films were measured.

The amount of non-freezing water of alginate with various kinds of ions was obtained by using the enthalpy of melting of water in each system [34]. The method for calculating the bound water content is found in 2.1.1 in Chapter 3.

The contribution of freezing bound water explained in 3.4 was ignored here, since the amount of freezing bound water is not significantly observed in these systems. Figure 4-6 shows non-freezing water content of NaAlg, and other alginates substituted by Ca, Ni, Ba, Cu, Al and Fe^{2+}. The linear relationship between non-freezing water content and ionic radius was not established in the above systems, although non-freezing water content of carboxymethylcellulose - mono- and di-valent cation systems decreased linearly with increasing ionic radius. Table 4-1 shows several examples of ion species and non-freezing water content.

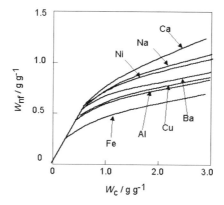

Figure 4-6. Relationship between non-freezing water content (W_{nf}) and W_c of various types of alginate. Sample preparation; Sodium alginate ($M_w = 9.5 \times 10^4$, degree of substitution 1.1, mannuronic acid and guluronic acid ratio 1.0 - 1.2. Sodium ion was converted into various types of ions in aqueous media. Measurements; Heat flux type DSC (Seiko Instruments), temperature range 120 to 320 K, heating rate =10 K min^{-1}, sample mass = 3 - 5 mg.

Table 4-1. Non-freezing water content (W_{nf}), degree of substitution (DS) and ionic radius of alginate with various counter ions

Ion	Ionic radius/nm	DS	$W_{nf} / g\ g^{-1}$
Na	11.6	1.0	0.82
Cu^{2+}	8.7	0.30	0.99
Fe^{2+}	7.5	0.31	0.98
Fe^{3+}	6.9	1.00	0.50

1.1.4   Ca alginate hydrogels

Figure 4-7 shows representative DSC cooling and heating curves of water in CaAlg gel fibres. Sample preparation of CaAlg gel fibres is found in the caption of Figure 4-7. When the peaks are magnified, it is seen that the shape of melting and crystallization peak of water becomes broader with decreasing water content. At the same time, the low temperature side sub-peaks become larger. It is noted that the melting patterns are quite similar to those observed in cellulosic hollow fibres [30].

By polarizing microscopy, molecular orientation can clearly be seen. Using a photo sensitive plate, colour change was confirmed when the fibre is moved 90 degrees. This indicates that the b-axis of molecular chains aligns to the fibre axis (Figure 4-8).

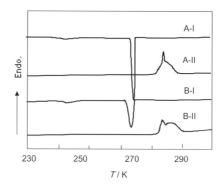

Figure 4-7. DSC curves of water in CaAlg fibres. A: concentration = 4 wt % (W_c=24 g g^{-1}), B; concentration = 5 % (W_c= 18 g g^{-1}), I: cooling, II: heating Samples; Gelation of NaAlg was carried out as follows; (1) 0.5, 1.0 and 2.0 wt % of NaAlg aqueous solution was prepared at 298 K. (2) 10 ml solution was put in a glass vessel with inner and outer lids. (3) a tightly sealed vessel was annealed at 378 K for 30 min in an oven, (4) aqueous solution of NaAlg was extracted in a microsyringe (5) the solution was injected into an excess amount of CaCl$_2$ aqueous solution of 3 wt % (6) transparent hydrogels were maintained in CaCl$_2$ aqueous solution for a predetermined time, (7) gel fibre was taken out and washed in water repeatedly (8) gel fibres were maintained in deionized water at 278 K. Measurements; power compensation type DSC (Perkin Elmer). scanning rate = 10 K min^{-1}, sample mass = ca. 3 mg.

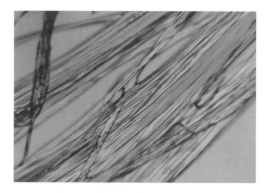

Figure 4-8. Polarizing light micrographs of CaAlg fibres in water.

As described in the initial part of this section, when NaAlg aqueous solution was dropped in CaCl$_2$ solution, spherical hydrogels were obtained. Preparation of hydrogels is explained in the caption of Figure 4-9. The rigidity of hydrogels was measured using a thermomechanical analyzer in water and it was confirmed that rigidity increases with increasing concentration of NaAlg aqueous solution and time of conversion in CaCl₂ solution. The practical application of this type of hydrogel is found elsewhere [35]

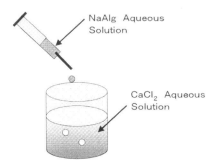

Figure 4-9. Schematic illustration of CaAlg hydrogels in sphere shape. Gelation of NaAlg was carried out as follows; (1) 0.5, 1.0 and 2.0 wt percent of NaAlg aqueous solution was prepared at 298 K. (2) 10 ml solution in a glass vessel with inner and outer lids. (3) a tightly sealed vessel was annealed at 378 K for 30 min in an oven. (4) then, the sample was maintained at 323, 333 and 343 K for 5, 30, 500 and 1000 min. (5) annealed samples were maintained at 298 K. (6) using a 1 ml syringe, the solution was dropped in an excess amount of CaCl₂ aqueous solution of 3 wt %. (7) spherical gel was immediately obtained (8) gels were removed from CaCl₂ solution, washed in water and kept in water for several days.

1.1.5   TMA of CaAlg gel films

When CaAlg hydrogels film are dried thin, transparent films are obtained. This gel film has a large water swelling capability. Swelling behaviour of hydrogels has been investigated by the measurement of mass and size change of the sample at static conditions as described in 2.2.3 (Chapter 2). In order to measure the swelling behaviour of gel films, thermomechanical measurement in water is a useful technique [36]. Using the sample holder shown in Figure 2-13 (Chapter 2), the dynamic swelling process of gel films can be investigated.

Figure 4-9 shows a representative swelling curve of NaAlg gel film. The degree of swelling shown in the figure is defined as

$$\varepsilon = \frac{(l_t - l_0)}{l_0} \times 100 , \% \tag{4.3}$$

where l_0 is thickness of dry gel film and l_0 is that of swollen film. As shown in Figure 4-10, the degree of swelling is ca. 4500 %. When CaAlg gel films were prepared by the sample preparation shown in the caption, CaAlg sheets pile up in the perpendicular direction. Water is retained between the layers and expansion is extremely large, in contrast when the degree of swelling is measured in the transverse direction, ε value was 30 %.

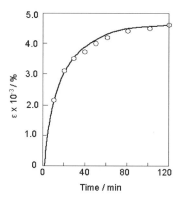

Figure 4-10. Representative swelling curve of NaAlg gel film. Sample preparation 5 wt % NaAlg (mannuronic and guluronic component ratio = 1.0) was solved in deionized water at 293 K and maintained for 24 hr. The dope was extended on a glass plate and water in the solution was evaporated. A partly dried sample on glass plate was immersed in an excess amount of 3 wt % of CaCl2 aqueous solution and maintained for several hours. After that, the sample was washed and water insoluble CaAlg gel film was removed from the plate. The thickness of completely dried films was 20 to 100 μm. Measurement; TMA (Seiko Instruments) column shaped quartz probe (Chapter 2, Figure 2 -10, A) with cross section 33 mm2, applied load 1 - 10 g, temperature 293 K sample size; diameter ca. 8 mm. Deionized water was used..

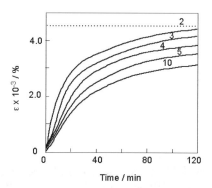

Figure 4-11. Swelling curves of CaAlg gel films under various loads. The numerals in the figure show loads (g). The broken line shows the degree of swelling obtained by static method. The films were immersed in deionized water for 120 min at 294 K. Measurements; TMA (Seiko Instruments), sample cell (see Chapter 2, Figure 212).

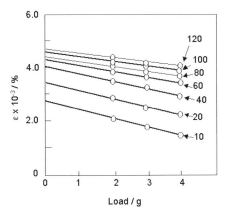

Figure 4-12. Relationships between degree of swelling (ε) and applied load of CaAlg gel film at various times of swelling. Numerals show swelling time (min).

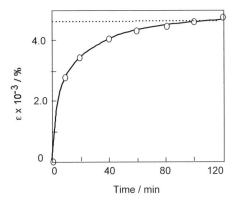

Figure 4-13. Swelling curve without load obtained from extrapolated values in Figure 4-11. The broken line shows the value obtained by static method.

When swelling behaviour is measured by TMA, a certain amount of load must be applied, since the quartz probe should have good contact with the surface of the samples. If the samples are soft and easily deformed by applied load like hydrogels, the results are necessarily affected by the amount of load. Figure 4-11 shows variation of swelling ratio under various loads as a function of time. When a load is small, swelling ratio reaches the value obtained by static measurement. Figure 4-12 shows relationships between degree of swelling (ε) and applied load of CaAlg gel film at various times of swelling. From linear relations, it is possible to extrapolate to zero loading. When the extrapolated values are related with time, swelling curve of CaAlg film without loading is obtained as shown in Figure 4-13. By TMA, not only the swelling behaviour but also creep and relaxation of hydogels in water can be measured

1.2   Gelation of xanthan gum by annealing and subsequent cooling

Xanthan gum is a polysaccharide produced by *Xanthomonas campestris*, which is a trisaccharide side-chain that attaches to alternate D-glucosyl residue (Figure 4-14). Due to the helical structure of xanthan gum [37, 38], it is assumed that rigid rod 5_1 double helix is formed in a dilute aqueous solution. It is also reported that rigid rod molecular structure forms the cholesteric liquid crystalline state in aqueous media [39].

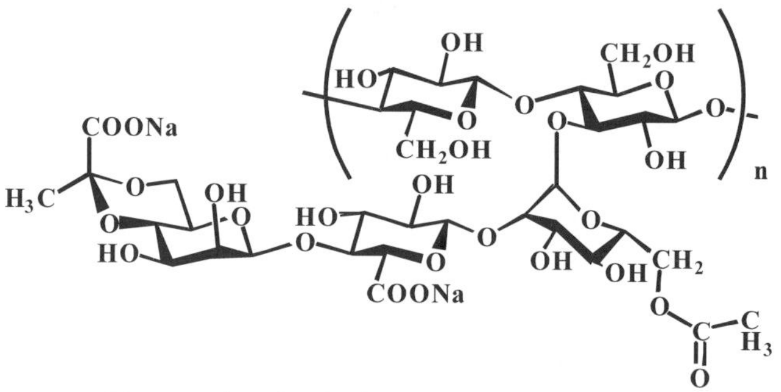

Figure 4-14. Chemical structure of xanthan gum.

Xanthan gum has been known as a non-gelling polysaccharide in aqueous solution. In order to prepare hydrogels, other polysaccharides have been mixed with xanthan gum [40]. However, it was found that xanthan gum forms hydrogels if aqueous solution is annealed at a temperature higher than the characteristic temperature and subsequently cooled [41, 42, 43]. By the falling ball method (see Chapter 2, 2.2.5) and viscometry, it was confirmed that the above characteristic temperature corresponds to the gel-sol transition temperature (T_{g-s}) [42]. Figure 4-15 shows the height of a steel ball suspended in xanthan hydrogels, prepared by annealing the solution in sol state and subsequent cooling, as a function of temperature. As shown in the figure, the steel ball fixed in the gel matrix starts to fall at around 310 K, indicating that the gel sol transition takes place.

Figure 4-16 shows gel sol transition temperatures of xanthan gum hydrogels prepared by annealing at 313 K for various time intervals. The gel sol transition temperatures of 2 wt % gels are higher than those of 1 wt%

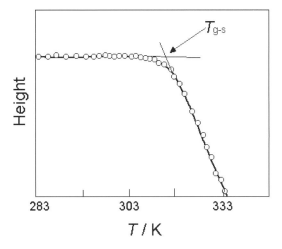

Figure 4-15. Relationship between the height of the steel ball and the temperature of 2 wt % xanthan hydrogel prepared by annealing and subsequent cooling. Temperature of annealing 313 K, time; 12 hr. Measurements; see for details 2.2.5 in Chapter 2, heating rate = 0.5 K min⁻¹, diameter of steel ball = 0.8 mm, mass of steel ball = ca. 0.5 mg.

gel in a whole range of annealing times. The temperatures level off at around 10 hours. The results indicate that gel-sol transition temperature depends not only on concentration but also on time of pre-annealing at the sol state. Temperature-time conversion is also applicable to the xanthan gum hydrogels.

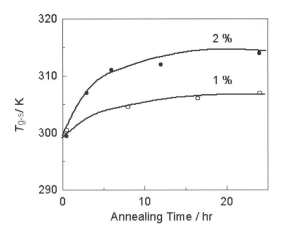

Figure 4-16. The gel sol transition temperature of xanthan gum hydrogels annealed at 313 K for various annealing times. Mesurement; Falling ball method (2.2.5).

Dynamic modulus of hydrogels prepared by pre-annealing was measured by TMA using an oscillation method (ref. 1.3, Chapter 2). Figure 4-17 shows dynamic modulus (E') of xanthan gum hydrogels prepared by annealing at various temperatures and subsequent cooling as a function of annealing time. E' increases in the initial stage of pre-annealing. Values of E' increase with increasing annealing temperature in the whole range of annealing time.

.

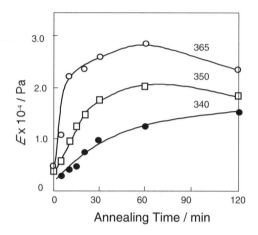

Figure 4-17. Dynamic modulus of xanthan gum hydrogels prepared by pre-annealing was measured by TMA using an oscillation method (ref. 1.3, Chapter 2).

Figure 4-16. shows dynamic modulus (E') of xanthan gum hydrogels prepared by annealing at various temperatures and subsequent cooling as a function of annealing time. E' increases in the initial stage of pre-annealing. Values of E' increase with increasing annealing temperature in the whole range of annealing time.

It is considered that gel-sol transition temperature and dynamic modulus vary when the number or size of the junction zone (cross-linked area) varies. In order to clarify the structural change of junction zone, xanthan gum hydrogels were measured by small angle x-ray diffractometry (SAXS) [41].

SAXS patterns of xanthan gum hydrogels were measured using synchrotron orbital radiation (SOR). Experimental conditions are shown in the caption of Figure 4-18. The position of the peak varies according to the annealing time. The long period d obtained from SAXS scattering peaks is shown in Figure 4-19. The position of the maximum of the synchrotron orbital radiation (SOR) scattering peak decreases in the initial stage, reaches a minimum and then increases with increasing annealing time. Similar behaviour of the distance of the long period was observed not only for xanthan gum hydrogels but also for gellan hydrogels obtained by pre-annealing and subsequent cooling.

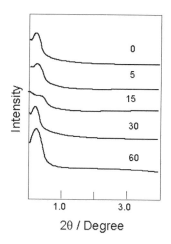

Figure 4-18. Changes of small angle x-ray profile of xanthan gum solution with concentration 2.0 wt % as a function of annealing time at 365 K. A diffraction peak is observed for 0.1 < 2θ < 0.4 degrees. Numerals in the figure show annealing time (min). Measurement; Synchrotron orbital radiation (Tsukuba, BL-10C beam line) was used. x-ray wavelength = 0.1488 nm, camera length = 2.0 M. sensitive proportional photon counter (PSPC), cross-referenced with standard collagen, was used. Temperature of measurement = 293 K.

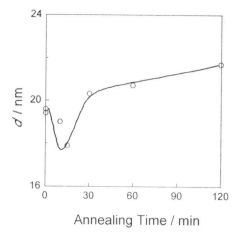

Figure 4-19. The long period *d* obtained from SAXS scattering peaks as a function of annealing time at 365 K for the 2 wt % xanthan gum solution.

The above experimental results strongly suggest that conformation of molecular chains of xanthan gum changes during annealing in sol state. When the amount of non-freezing water is measured during annealing as a function of time, the amount fluctuates in an oscillatory manner as shown in Figure 4-19. It is thought that the number of non-freezing water molecules restrained by xanthan gum necessarily reflects the conformational structure.

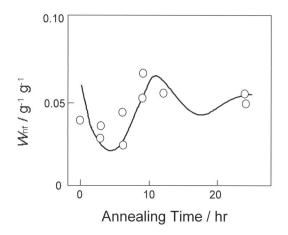

Figure 4-20. Comparison between experimental data and calculated values using equation 4.13.

In order to explain the behaviour of non-freezing water in xanthan gum-water systems, the following model was suggested [44 - 47] Assuming the total number of repeating units of polysaccharide molecules (N) is defined as equation 4.4

$$N = N_0 + N_u + N_s \tag{4.4}$$

where N_o is the number of repeating units which adsorbed water molecules, Nu those not adsorbing water molecules and N_s those being incapable of adsorbing water molecules. If s-monomers are needed to adsorb one water molecule, equation 4-5 is obtained

$$N_0 = sn,$$
$$N_u = sn_u \tag{4.5}$$

where n is the number of water molecules adsorbed on the polysaccharide molecular assemblies and n_u is the number of adsorption sites unoccupied by water molecules. It is thought that n increases when n_u is sufficiently large. The time evolution of n is expected to obey

$$\frac{dn}{dt} = k_1 n_u + k_2 n \tag{4.6}$$

where k_1 and k_2 are rate constants for adsorption and desorption of water molecules. According to equation 4.4, n changes with time t, if n deviates from the equilibrium value n_{eq}. The system is in the non-equilibrium state and equations from 4.4 to 4.6 give

$$\frac{dn}{dt} + \left(k_1 + k_2\right)n = \frac{k_1}{s}\left(N - N_s\right) \tag{4.8}$$

With annealing time t, $\frac{dn}{dt}\left(k_1 + k_2\right)n$, is thought to change. If the simple relation to describe the time evolution of N-Ns is assumed,

$$\frac{d(N - N_s)}{dt} = -\alpha\left[n - n_{eq}(T)\right] \tag{4.8}$$

Here, α is the phenomenological constant and n_{eq} (T) is the equilibrium value of n at the annealing temperature T. From equations 4.8 and 4.9, the differential equation is obtained.

$$\frac{d^2 n}{dt^2} + \left(k_1 + k_2\right)\frac{dn}{dt} + \frac{k_1 \alpha n}{s} = \frac{k_1 \alpha n_{eq}}{s} \tag{4.10}$$

New parameters γ and ω and the new variable n are redefined as follows,

$$\gamma = \frac{\left(k_1 + k_2\right)}{2}$$
$$\frac{k_1 \alpha}{s} = \omega_0^{\ 2} \tag{4.11}$$
$$n = n - n_{eq}$$

Here, γ represents the averaged rate constant for adsorption and desorption of water molecules and ω_0 the constant representing the effect of the change of the number of adsorption sites on the time evolution of n. Equation 4-10 is transformed by the replacement of equation 4-11 and the equation 4-12 is obtained.

$$\frac{d^2n}{dt^2} + \frac{2\gamma dn}{dt} + \omega_0^2 n = 0 \tag{4.12}$$

In the case of $\gamma < \omega_0$, equation 4.12 has the oscillational solution

$$n = C\exp(-\gamma t)\cos(\omega t + \delta) \tag{4.13}$$

where $\omega = \left(\omega_0^2 - \gamma^2\right)^{\frac{1}{2}}$. The solid line shown in Figure 4-19 is the theoretical curve calculated by equation 4.13 [44].

1.3   Sol-gel transition of curdlan

Curdlan (Figure 4-21) is a unique water insoluble polysaccharide. Curdlan forms hydrogel when aqueous suspension is annealed at a temperature higher than characteristic temperature [48, 49]. It is reported that curdlan hydrogels become thermo-irreversible when curdlan suspension is annealed at a temperature higher than ca. 60 °C. This type of gel is designated as high set gels [50]. When the annealing temperature is lower than 60 °C, gels are thermoreversible and are designated as low set gels. Methylcellulose (MC) is also known as a gel forming polysaccharide by heating. MC is water soluble and forms hydrogels by elevating temperature. MC hydrogels are thermoreversible and transform to the sol state by cooling (see section 5.2 of chapter 3). Sol-gel and gel-sol transition of MC are clearly seen even by visual observation. In contrast, gelling conditions of curdlan aqueous suspension are more complex. By x-ray diffractometry,

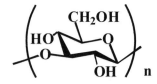

Figure 4-21. Chemical structure of curdlan.

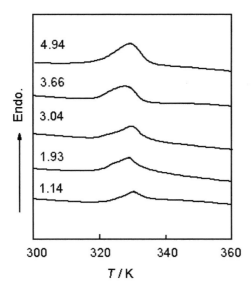

Figure 4-22. DSC heating curves of curdlan suspensions. Numerals in the figure show concentration (wt %.). Measurements; high sensitivity DSC (Seiko Instruments) with a cooling attachment, sealed type silver pan (70 μl), sample mass = 50 μl, heating rate = 1 K min^{-1}, N$_2$ flow rate = 10 ml min^{-1}.

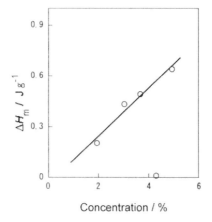

Figure 4-23. Relationship between enthalpy of transition and concentration of curdlan.

curdlan molecules form triple helices in aqueous media [50]. The higher–order structure of curdlan in various conditions such as the presence of cations has been reported by many authors [51, 52, 53]. By thermoanalytical measurements [51, 53], an endothermic peak was observed which is thought to be a sol-gel transition.

Figure 4-22 shows DSC heating curves of curdlan suspensions with various concentrations. An endothermic peak due to gel-sol transition is observed. Transition temperature maintains a constant value. Enthalpy of transition increases linearly with increasing concentration as shown in Figure 4-23.

1.4   Freezing and thawing gelation of locust bean gum

Among various types of polymers having gel forming capabilities, poly(vinyl alcohol) (PVA) has been reported to form hydrogels by freezing and thawing [54, 55, 56]. It is also known that locust bean gum (LBG) forms hydrogels by freezing and thawing [57, 58]. If the formation and dissociation of polysaccharide gels are controllable, the water holding and releasing function can be programmed. On this account, gels prepared by freezing and thawing are considered to be material whose dissociation of crosslinking can be manipulated. LBG is a polysaccharide whose chemical structure is shown in Figure 4-24. 1,6 α-D- galactose side chain is attached to each 4 repeating units of 1, 4-β-D-mannose main chain. When the side chain is attached to either three (tara gum) or two repeating units (guar gum) [59, 60], gelation is not observed. It is thought that an appropriate chemical structure to form hydrogen bonding between the hydroxyl groups is crucial in order to form gels. At the same time, it is found that gelation by freezing and thawing is affected by cooling rate which affects the size of ice formed in the LBG solution [58]. It is thought that the size of ice necessarily corresponds to the size of the junction zone which is known as the crosslinking point of gels.

When the number of freezing and thawing cycles (n) of aqueous solution of LBG increased, the rigidity of the gels increased. When the concentration is higher than 1 %, the amount of sol transformed to gels was 100 % regardless of n. On the other hand, when concentration is less than 1 wt %, gel ratio (= (mass of gel) / (total mass of solution), g g^{-1}) decreases with increasing n, suggesting that the separated sol portion increases.

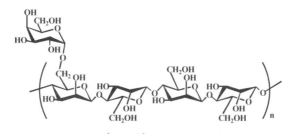

Figure 4-24. Chemical structure of locust bean gum.

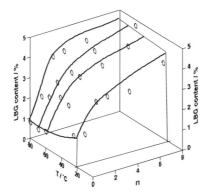

Figure 4-25. Relationships between LBG content in the gel portion, number of freezing and thawing cycles (*n*) and annealing temperature of the sample in the sol state (1 wt %)Sample preparation; (1) LBG aqueous solutions (1wt %) were prepared in a glass vessel with a polyethylene inner lid and a plastic outer lid (2) each vessel was maintained at 378 K for 2 hours. No mass loss was observed during heating at 378 K. (3) After annealing at 378 K, the first series of samples were directly transferred to a freezer whose temperature was 258 K and maintained for 24 hours. The second series of samples were cooled to 378 K maintained for 2 hours and then frozen at 258 K. The third series of samples were annealed at 333, 343 and 353 K for 2 hours then frozen at 258 K. (4) Frozen samples were thawed at 378 K. This process took more than 6 hours. (5) The above freezing and thawing was repeated from 1 to 10 times. In this study, the number of freezing and thawing cycles is stated as "*n*". After freezing and thawing, the samples were stored at 378 K.

LBG hydrogels formed by freezing and thawing are thermo-irreversible. Once the junction zone is established, the resolution of molecular chains does not occur even if the gel is maintained at a temperature higher than 330 K. This fact is quite different from that of PVA hydrogels prepared by freezing and thawing which are completely thermo-reversible. As described in 1.2 of this chapter, the structure of xanthan gum in the sol state changes by equilibration at a temperature higher than sol-gel transition. In the case of LBG, a similar experiment was carried out. Figure 4-25 shows relationships between LBG content in the gel portion, number of freezing and thawing cycles and annealing temperature of the sample in the sol state. The effect of annealing LBG sol is clearly seen.

The amount of water tightly restrained by the junction zone of LBG hydrogels was calculated using enthalpy of melting obtained by DSC. It is known that water molecules which are strongly restrained by matrix polysaccharides show no first order phase transition. This kind of water is defined as non-freezing water (W_{nf}). W_{nf} values evaluated from DSC melting peak of LBG hydrogels as functions of freezing and thawing cycles(*n*) are shown in Figure 4-25. As shown in this figure, the W_{nf} value decreases with increasing *n*. This fact suggests that LBG molecules in the junction zone become regularly aligned, and water molecules directly bonded to hydroxyl groups of LBG are excluded when inter-molecular hydrogen bonding is established.

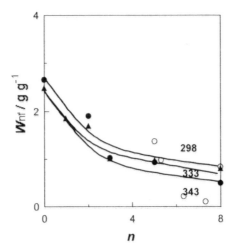

Figure 4-26. Relationships between non-freezing water content of LBG hydrogels and number of freezing and thawing cycles *(n)*. Numerals in the figure show annealing temperature of LBG sol (1 wt %).

From the above results, it is concluded that the junction zone of LBG hydrogels is tightly formed by freezing and thawing. During this process, LBG molecules not connected with the junction zone are excluded from the gel portion and remaining molecules gradually form densely packed hydrogels. Molecular conformation in the sol state affects the rate of the above junction formation. This fact suggests that LBG solution is not in thermally equilibrium state. Obtained LBG hydrogels are thermally stable and no gel-sol transition was observed in a temperature ranging from 310 to 373 K. Non-freezing water content calculated from DSC melting peak of water in the gel indicates that the junction zone becomes dense with increasing the number of freezing and thawing cycles since water restrained in the junction zone is excluded by the freezing process.

1.5   Effect of annealing on gelation of pectin

Pectin is a polysaccharide found in many plant tissues, such as fruits and vegetables [61]. For example, pectin content of apple is 0.5 - 1.6, grape 0.2 - 1.0, orange peel 3.5 to 5.5 lemon 3.0 -4.0 and carrot 1.0 wt %, respectively. The principal constituent of pectin is D-galacturonic acid, joined in chains by α(1-4) glycosidic linkages. Pectin molecular chains contain a part of the carboxyl groups esterified by methyl group [62, 63]. The degree of methyl esterification (DE) ranges widely according to the kinds of plant. Pectin is sometimes designated as pectinic acid. Pectin without methyl esterification is called pectic acid. The conformational structure of pectins and pectic acid is found elsewhere [63].

Pectinic acid is water insoluble. Water solubility of pectins depends on DE and degree of polymerization [64]. Low methoxyl pectins form gels in the presence of divalent cations, in contrast high methoxy pectins form gels in the presence of sugars in acidic conditions. Figure 4-27 summarizes the solubility and gelation of pectins and pectic acid.

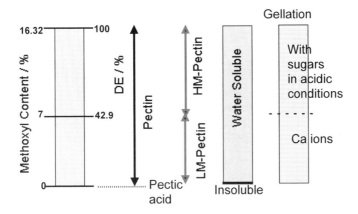

Figure 4-27. Solubility and gelation of pectins and pectic acid (Iijima M., Thermal properties of pectin hydrogels, Doctoral Thesis, Otsuma Women's Univerity, 2001). DE; degree of esterification.

Similar to other polysaccharide sols, when pectin hydrogels are prepared in the presence of Ca ions after aqueous solution is sufficiently equilibrated, it was found that physical properties of hydrogels remarkably depend on annealing conditions in the sol state [65]. When pectin sol was annealed at high temperature, Ca pectin formed softer gels than those annealed at low temperature. On the other hand, in swelling measurements by tea bag method, the swelling ratio decreased with increasing annealing temperature at sol state. The non-equilibrium state of pectin sol can be fixed by Ca ions.

Mechanical properties of hydrogels have mainly been investigated by viscoelastic measurement using a rheometer. Even when the gel is rigid, mechanical testing of hydrogels using a custom made machine is limited, since the gel frame is frequently destroyed by clamping. Furthermore, vaporization of water during measurements is ignored. In this section, the application of a thermomechanical analyser (TMA) to measure dynamic modulus of hydrogels in water is described [66]. TMA enables us to measure the static and dynamic mechanical properties of hydrogels in water [66].

Table 4-2. Characteristics of pectin samples

Number	DE/%	M_w
A	25.6	8.5×10^4
B	29.6	8.5×10^4
C	60.7	1.0×10^5
D	71.9	1.5×10^5

Pectin samples were in powder form, whose DE and molecular mass are shown in Table 4-2. Pectin powder was solved in deionized water at 298 K and 2 wt % solution was obtained. The solutions were stirred and annealed at 298, 323, 353 an 371 K for 30, 80 and 180 min, respectively. After the solution was equilibrated at 298 K, an aqueous solution of CaCl$_2$ (0.43- 3.42 %) was poured into a pectin solution. Transparent flexible gels were immediately formed. The pectin concentration of samples at this state was 1.0 to 1.18 %. The amount of Ca ions was measured by atomic absorption spectrometry and degree of substitution (DS) was calculated. DS is defined as indicated in equation 4.14.

$$DS = \left(\frac{Ca_{meas}}{Ca_{cal}} \right) (1 - DE) \qquad\qquad (4.14)$$

where Ca$_{meas}$ is measured concentration of Ca ion in the gel, Ca$_{cal}$ calculated concentration assuming all COOH is substituted by Ca ions and DE is degree of esterification. DE increased with increasing CaCl$_2$ concentration and leveled off at a characteristic value which depends on the annealing condition of pectin sol. DS decreased with increasing annealing temperature and time at sol state. Swelling ratio calculated using the results obtained by tea bag method (see 2.3.3 Chapter 2) increased with decreasing DS.

Dynamic modulus and dynamic loss modulus of Ca pectin hydrogels were measured by TMA using the experimental procedure shown in 1.3 in Chapter 2. Figure 4-28 shows relationships between E', tan δ and degree of substitution of Ca ions (DS). *E'* increases linearly with increasing DS. The effect of annealing is not significantly observed by TMA.

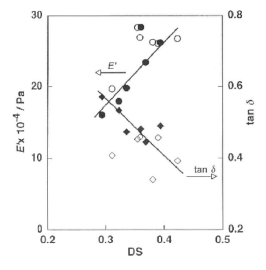

Figure 4-28. Relationships between dynamic modulus (*E'*), tan δ and degree of substitution (DS). J, F; DE = 29.6 % , E, p ; DE = 25.6 %.

2.   GLASS TRANSITION AND LIQUID CRYSTAL TRANSITION

Numerous polysacchrides form hydrogels in aqueous media whether they are electrolytes or non-electrolytes [67, 68] In this section lyotropic and thermotropic liquid crystallization of several polysaccharides is described.

2.1   Glass transition and liquid crystal transition of xanthan gum-water systems

As described in section 1.2 of this chapter, the persistence length of xanthan gum is long, and hence, molecular chains readily align in the presence of a certain amount of water. Similar to sodium cellulose sulfate described in chapter 3, liquid crystal formation of Na xanthan gum-water systems depends markedly on water content and temperature. At the same time, structural change of water in the systems was clearly observed in relation to liquid crystal formation [69]. In this section, liquid crystal formation and glass transition of xanthan gum-water systems are described

Figure 4-29 shows representative DSC heating curves of xanthan gum-water systems. In the system with W_c 0.24 g g^{-1}, a distinct glass transition,

cold crystallization, small melting peak of water and large transition peak
from liquid crystal to liquid state are observed. When W_c increases to 0.84 g
g^{-1}, cold crystallization is hardly observed, melting peak increases and
transition peak from the liquid crystal to liquid state becomes smaller. In the
system with W_c 1.40 g g^{-1}, cold crystallization and transition peak from
liquid crystal to liquid state disappear.

A phase diagram is established as shown in Figure 4-30, using glass
transition temperature (T_g), peak temperatures of cold crystallization (T_{cc}),
melting peak of water (T_m) and transition peak from liquid crystal to liquid
state (T^*). As clearly seen in Figure 4.28, ice in the system is formed at a
characteristic W_c where T_g value is at a minimum value. When T_m levels off,
T_{cc} disappears. T^* decreases with increasing W_c and disappears when T_g and
T_m level off.

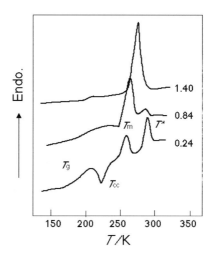

Figure 4-29. Representative DSC heating curves of xanthan gum-water systems. Numerals
show water content (W_c).Samples; degree of substitution 2.6, Experimental conditions; power
compensation type DSC (Perkin Elmer), heating rate =10 K min^{-1}.

The amount of non-freezing water (W_{nf}) was calculated using transition
enthalpies obtained shown in Figure 4-31, Figure 4-31 shows the
relationship between non-freezing water content (W_{nf}), freezing water
content (W_f) and water content (W_c) of xanthan gum - water systems. The
maximum amount of W_{nf} is ca. 5.1 g g^{-1}. The amount coincides with the W_c

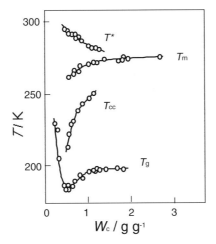

Figure 4-30. Phase diagram of xanthan gum - water systems. T_g; glass transition, temperature, T_{cc}; cold crystallization, T_m; melting of water, T^*; transition peak from liquid crystal to liquid state. Experimental conditions, see Figure 4.2 caption.

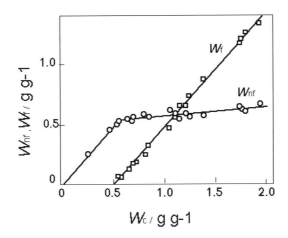

Figure 4-31. Relationships between non-freezing water content (W_{nf}), freezing water content (W_f) and water content (W_c) of xanthan gum-water systems.

where T_g showed the minimum value. When W_c exceeds 0.51 g g^{-1}, ice is formed in the system and cooperative molecular motion of xanthan gum and non-freezing water is restricted. At the same time, molecular chains align more easily due to the existence of movable water, hence the liquid crystalline phase is formed. Figure 4-32 shows polarizing light micrographs of xanthan gum - water system with W_c = ca. 1.0 g g^{-1}. As shown in Figures 4-28 and 4-31, both transition temperature (T^*) and transition enthalpy decrease with increasing W_c. This suggests that molecular mobility is enhanced and the ordered structure is difficult to maintain when an excess amount of free water exists in the system.

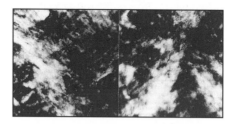

Figure 4-32. Polarizing light micrographs show liquid crystalline phase of xanthan gum-water systems. Temperature =298 K. W_c = ca. 1.0 g g^{-1}. A Leitz polarizing microscope, Orthoplan POL, photo sensitive plate was used.

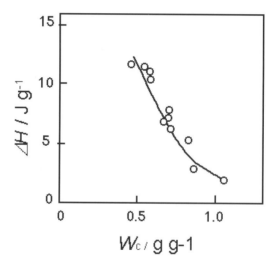

Figure 4-33. Relationship between transition enthalpy from liquid crystal to liquid phase and water content (W_c) of xanthan gum-water systems.

Both T_g value and heat capacity gap at T_g (ΔC_p) can be used as a criteria of molecular mobility of xanthan gum-water systems. ΔC_p reflects the randomness of molecular chains. Since glass transition of the system is a molecular enhancement due to cooperative motion of xanthan gum chains associating with non-freezing water, the molecular order of xanthan gum chains plays a crucial role in ΔC_p values [70]. Figure 4.32 shows DSC curves of xanthan gum-water system with W_c = 0.7 g g^{-1}. As shown in Figure 4.32, ΔC_p value observed in DSC heating curve (A) of the system cooled from the liquid state is larger than that (B) of the system cooled from the liquid crystalline state. In these two heating curves, T_g, T_{cc}, T_m and T^* are observed at a similar temperature. This strongly suggests that the structure formed at the liquid crystalline state is frozen at a temperature lower than T_g.

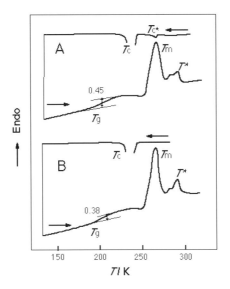

Figure 4-34. DSC curves of xanthan gum-water system with different thermal histories. W_c = 0.7 g g-1, heating rate = 10 K min^{-1}.

2.2 Phase diagram of three galactomannans, locust bean gum, tara gum and guar gum

Locust bean gum (LBG), tara gum (Tara-G) and guar gum (GG) are categorized as galactomannan polysaccharides which consist of a 1,4-β-D-mannose backbone and 1,6-α-D-galactose side chains. A structural difference among the three gums is the galactose / mannose ratio, which is 1:4, 1:3 and 1:2 for LBG, Tara-G and GG, respectively. These chemical structures are shown in Figure 1. Gelation of the above three galactomannan polysaccharides in aqueous systems has been carried out by chemical modifications, in the presence of cross linking agents, and mixing with other polysaccharides, such as xanthan gum and conjac mannan. Gellation of LBG by freezing and thawing has already been described in section 1.4 in this chapter. It is thought that regularly spaced side chains have a crucial role in gel forming ability. In this section, the phase transition behaviour of a series of galactomannan polysaccharide-water systems was investigated using differential scanning calorimetry (DSC).

Phase transition behaviour of LBG-, tara- and GG-water systems with water content from 0 to ca. 5.0 g g^{-1} was investigated by DSC [71]. Using phase transition temperatures obtained from DSC heating curves, phase diagrams were established. As shown in Figures 4-34, 4-35 and 4-36, T_g, T_{cc} and T_m are observed in all of the three samples. T_g and T_{cc} are found in a W_c range from 0.5 g g^{-1} to 1.0 g g^{-1} for LBG and Tara gum-water systems (Figures 4-34 and 4-35). On the other hand, in GG-water systems, T_g and T_{cc} are observed in a wide W_c range from 0.5 g g^{-1} to 2.7 g g^{-1} (Figure 4-36). As shown in Figures 4-34, 4-35 and 4-36, T_m is observed when W_c exceeds 0.4 g g^{-1} for all samples. T_m increased with increasing W_c, and leveled off at around $W_c = 1.0$ to 1.2 g g^{-1}. Leveling off temperatures of the three samples are observed at 273 K. In the case of pure water, peak temperature of melting was observed at 278.4 K at the heating rate 10K min^{-1} due to super-heating. The above facts suggest that freezing water coexists with a considerable amount of freezing bound water in these galactomannan-water systems.

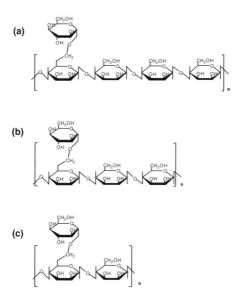

Figure 4-35. Chemical structure of locust bean gum (LBG) (a), tara gum (tara-G) (b) and guar gum (GG)(c).

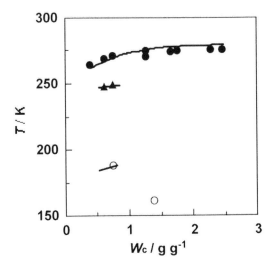

Figure 4-36. Phase diagram of locust bean gum – water systems. T_m; melting peak temperature, T_{cc}: cold crystallization peak temperature, T_g; glass transition temperature Measurement; heat flux type DSC (Seiko Instruments), heating rate = 10 K min^{-1}, N$_2$ flow rate = 30 ml min-1, aluminium sealed type pan, sample mass = ca. 3 mg.

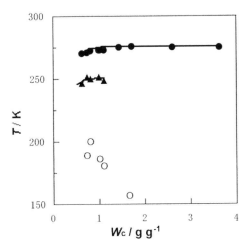

Figure 4-37. Phase diagram of tara gum – water systems. T_m; melting peak temperature, T_{cc}: cold crystallization peak temperature, T_g glass transition temperature.

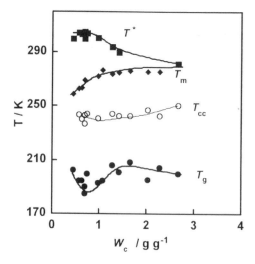

Figure 4-38. Phase diagram of guar gum – water systems. T^*; transition temperature from the liquid crystalline state to liquid state, T_m; melting peak temperature, T_{cc}: cold crystallization peak temperature, T_g; glass transition temperature.

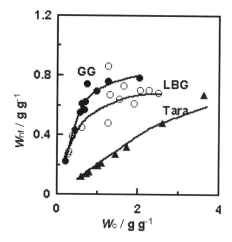

Figure 4-39. Relationships between non-freezing water content (W_{nf}) and water content (W_c) of LBG, tara gum and guar gum (GG).

In order to understand the structural change of water, the amount of bound water content was calculated based on the enthalpy of melting of water in the systems. Water molecules coexisting with polysaccharides are categorized into three types, such as free water (W_f), freezing bound water (W_{fb}), and non-freezing water (W_{nf}). The characteristic of each type of water is that melting of W_f starts at 273 K, W_{fb} lower than 273 K, and W_{nf} shows no first-order transition. Figure 4-37 shows W_{nf} of three different galacto-annans. As described in section 1.4 of this chapter, LBG-water systems form gels by freezing and thawing process [59], and GG-water systems form liquid crystals as shown in the phase diagram (Figure 4-38). It is considered that the amount of W_{nf} is related to the gelation and liquid crystallization.

2.3 Amorphous ice and liquid crystal of guar gum-water

Guar gum (GG) is a member of the galactomannan family [1,2] consisting of 1,4-β-D-mannose backbone and 1,6-α-D-galactose side chains, and the galactose / mannose ratio is 1:2. GG has received particular attention as an industrial polysaccharide due to its rheological properties [78]. GG has been mixed with various kinds of polysaccharide in order to control the viscoelastic properties. At the same time, it is known that GG-polysaccharide mixtures form hydrogels when they are crosslinked by borate [79].

When GG-water systems with a W_c range from 0 to 3.0 g g^{-1} were measured by DSC in a temperature range from 120 to 320 K, heating and cooling curves of the system varied in a complex manner as a function of W_c. From the low to high temperature side, glass transition, cold crystallization, melting and transition from liquid crystallization to liquid state were observed in GG-water systems. Representative heating curves of GG-water systems with various water contents scanned at 10 K min^{-1} are shown in Figure 4-38. All samples shown in this figure were cooled from 333K to 120 K at a cooling rate of 10 K min^{-1}. When W_c is smaller than 0.7 g g^{-1}, the melting peak is smaller than that of the transition from liquid crystallization to liquid state. Three GG-water systems showing characteristic DSC heating curves were chosen, i.e. systems with W_c = 0.59, 0.81 and 1.71 g g^{-1}. The system with W_c = 0.59 g g^{-1} has a larger transition peak from liquid crystal to liquid state than that of melting; the system with W_c = 0.81 g g^{-1} peak has similar size; the system with W_c = 0.71 g g^{-1} has a large melting peak. The systems were cooled at cooling rates 1 or 10 k min^{-1} from the liquid state to glassy state and then heated at 10 K min-1. Figures 4-41, 4-42 and 4-43 show representative DSC heating curves of samples with W_c of 0.59, 0.81 and 1.71 g g^{-1} having different cooling histories.

The comparison was made among ΔH_c^*, ΔH_{cc} and ΔH_m^* for the sample with W_c =0.59 g g^{-1}. Figure 4-42 shows relationships between ($\Delta H_c^* + \Delta H_{cc}$) and ΔH_m^* as a function of cooling rate. It is clearly seen that ($\Delta H_c^* + \Delta H_{cc}$) vary in a similar manner to ΔH_m^*. This fact strongly indicates that liquid crystallization occurs at cold crystallization.

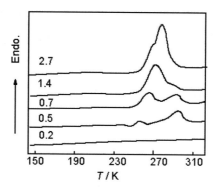

Figure 4-40. DSC heating curves of water-guar gum systems. Numerals shown in the figure show water content (g g⁻¹), Sample preparation. A small amount of water was added to dry powder and residual water was evaporated until a predetermined amount of water in the system was attained. Measurement conditions. heat-flux type DSC (Seiko Instruments) Heating rate 10 K min⁻¹. N_2 gas flow rate = 30 ml min⁻¹, Sample mass = ca. 5 mg, aluminium sealed type sample pan was used.

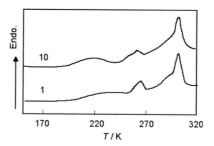

Figure 4-41. DSC heating curves of GG-water systems with W_c=0.59 g g⁻¹ cooled at 1 and 10 K min⁻¹. Measurements. See Figure 4-40 caption.

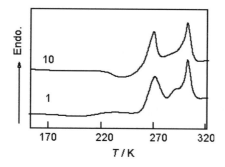

Figure 4-42. DSC heating curves of GG-water systems with W_c=0.76 g g⁻¹ cooled at 1and 10 K min-1. Measurements. See Figure 4-40 caption.

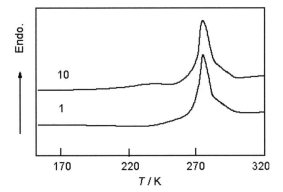

Figure 4-43. DSC heating curves of GG-water systems with W_c =1.71 g g^{-1} cooled at 1, 10 and 40 K min^{-1}. Measurements. See Figure 4-40 caption.

From the above results, it is concluded that the structural change of water in the GG-water system governs the liquid crystallization of the system. It is considered that lyotropic and thermotropic liquid crystals generally observed in water-polysaccharide systems are organized in the same manner as that of the GG-water system.

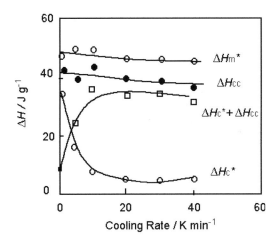

Figure 4-44. Relationship between enthalpy of transitions and cooling rate of GG-water systems with W_c=0.59 g g^{-1}. $\Delta H_c{}^*$; transition enthalpy from liquid to liquid crystalline state, ΔH_{cc}; transition enthalpy of cold crystallization, $\Delta H_m{}^*$; transition enthalpy from liquid crystal to liquid state [71].

REFERENCES

1.   Davidson, R. L. 1980, *Handbook of Water-soluble Gums and Resins*, McGraw-Hill book Co., New York.
2.   Daspinall, G. O. ed. 1982, The Polysaccharides, vol 1, Academic press, New York.
3.   Glass, J. E. ed. 1986, *Water-Soluble Polymers, Beauty with Performance*, Advances in chemistry series, ACS, Washington DS.
4.   Marchessault, R. H., 1984 *Carbohydrate Polymers, Nature's High Performance Materials, in Contemporary Topics in Polymer Science*, eds Vandernberg, E. J., Plenum Press, New York, pp. 15-53.
5.   Atkins, E. D. T. Ed. 1985, *Polysaccharides, Topics in Structure and Morphology*, VCH,. Weinheim.
6.   Guenet, J-M., 1992, Thermoreversible gelation of polymers and biopolymers, Academic Press, London.
7.   Horton, D., 1997, *Advances in Carbohydrate Chemistry and Biochemistry*, Academic press, San Diego.
8.   Hatakeyama, H. and Hatakeyama, T., 1998, Interaction between water and hydrophilic polymers (Review), *Thermochimica Acta*, **308**, 3-22.
9.   Hatakeyama, T., Nakamura, K. and Hatakeyama, H., 1996, Glass transition of the polysaccharide-water system", *Kobunshi Ronbunsyu*, **53**,795 (Japanese).
10. Watase, M., Nishinari, K., 1986, Rheology, DSC and volume or weight change induced by immersion in solvents for agarose and kappa-carrageenan gels, *Polym. J.,* **18**, 1017-1025.
11. Dea, I.C. M., McKinnon, A. A. and Rees, D. A., 1972, Tertiary and quaternary structure in aqueous polysaccharide systems which model cell wall cohesion; reversible change in conformation and association of agarose, carrageenan and galactomannans. *J. Mol. Biol.*, **68**, 153-172.
12. Rochas, C. and Rinaudo, M., 1984, Mechanism of gel formation in κ-carrageenan, *Biopolymer*, **23**, 735-75.
13. Thành, T. T. T., Yuguchi Y., Mimura, M., Yasunaga, H. Takano, R., Urakawa, H., Kajiwara, K., 2002, Molecular characteristics and gelling properties of the carrageenan family, 1. Preparation of novel carrageenans and their dilute solution properties, *Macromol. Chem. Phys.*, **203**, 1-23.
14. Yuguchi Y., Thành, T. T. T. Urakawa, H. and Kajiwara, K., 2002, Structural characteristics of carrageenan gels: temperature and concentration dependence. *Food Hydrocolloids*, **16**, 515-522.
15. Tanaka, R., Hatakeyama T., Hatakeyama H. and Phillips, G.O., 1996, Differential scanning calorimetric studies of Philippines natural grade κ-carrageenan. *Food Hydrocolloids,* **10**, 441-444.
16. Nishinari, K. ed., 1999, *Progress in Colloid Polymer Science Physical Chemistry and Industrial Application of Gellan Gum*, Springer-verlag, Berlin.
17. Ogawa, E., 1996, Conformational transition of polysaccharide sodium-gellan gum in aqueous solutions, *Macromolecules*, **29**, 5178-5182.
18. Quinn, F. X., Hatakeyama, T. Takahashi, M., Yoshida, H. and Hatakeyema, H., 1993, The conformational properties of gellan gum hydrogels, *Gels and Network*, **1**, 93-114.
19. Hatakeyama, T., Quinn, F. X., Hatakeyama, H., 1996, Changes in freezing bound water in water-gellan systems with structure formation, *Carbohydrate Polym.*, **30**, 155-160.
20. Hatakeyama, T., Nakamura, K., Takahashi, M. and Hatakeyama, H., 1999, Phase transition of gellan-water systems, *Progr. Colloid Polym. Sci.*, **114**, 98-101.

21. Zhang, H., Yoshimura, M., Nishinari, K., Williams, M. A. K., Foster, T. J. and Norton, I. T., 2001, Gelation behaviour of konjac glucomannan with different molecular weights, Biopolymer, 59, 38-59.

22. Dave, V., Stetah M., MaCarthy S. P., Ratto, J. A., Kaplan, D. L., 1998 Liquid crystalline, rheological and thermal properties of konjac glucomannan, *Polymer,* **39**, 1139-1148.

23. Ross–Murphy, S. B. and McEvoy H., Fundamentals of hyrogels and gelation, British Polym. J., 18, 1986 2-7.

24. Nishinari K., Zhang, H. and Ikeda, S., 2000, Hydrocoloid gels of polysaccharide and proteins, Current Option in Colloid & Interface Science, 5, 1995-2001.

25. Nishinari K., 1997, Rheilogical and DCS study of sol-gel transition in aqueous dispersions of indutrially important polymers and colloids, *Colloid Polym. Sci.*, **275**, 1093-1107.

26. Atkins, E. D. T., Nieduszynski IA., Wills H. H., Mackie, W., Parker, K. D. and Smolko, E. E., 1973, Structural components of alginic acid. I. The crystalline structure of poly-β-D-mannuronic acid. Results of x-ray diffraction and polarized infrared studies, *Biopolym.*, **12**, 1865-1878.

27. Atkins, E. D. T., Nieduszynski IA., Wills H. H., Mackie, W., Parker, K. D. and Smolko, E. E., 1973, Structural components of alginic acid. II. The crystalline structure of poly-α-LD-mannuronic acid. Results of x-ray diffraction and polarized infrared studies, *Biopolym.*, **12**, 1879-1887.

28. Larsen B. and Haug A., 1971, Biosynthesis of alginate, Part I. Composition and structure of alginate produced by *Azotobacter vinelandii* (Lipman), *Carbohyd. Res.*, **17**, 287-296.

29. Nakamura, K., Hatakeyama T. and Hatakeyama; H., 1993, Effect of Ca ions on the structural change of sodium alginate hydrogels"; in "*Cellulosic: Chemical, Biochemical and Material Aspects,* (J. F. Kennedy, G. O. Phillips and P. A. Williams eds.), Ellis Horwood, Chichester, p.243.

30. Quinn, F. X. Hatakeyama T. and Hatakeyama, H., 1994, Structural Change of Alginate Hydrogel Fibres in the Presence of Water, *Sen-i Gakkai-shi,* **50**, 418-420.

31. Nakamura, K., Hatakeyama T. and Hatakeyama; H., 1991, Formation of the liquid Ccrystalline state in the water-sodium alginate system, *Sen-i Gakkai-shi,* 47, 421-423.

32. Nakamura; K., Nishimura, Y., Hatakeyama, T. and Hatakeyama H., 1995, Thermal properties of water water content of mono- and divalent cation salts of polyelectrolytes-water systems studied by DSC, *Thermochimica Acta,* **253**, 343-353.

33. Takahashi, M., Kawasaki, Y., Hatakeyama T. and Hatakeyama, H., 2001, Effect of water on molecular motion of alginic acid having various guluronic and mannuronic acid contents, in "Recent advances in environmentally compatible polymers", (J. F. Kennedy, G. O. Phillips, P. A. Williams and H. Hatakeyama eds.), Woodhead Publishing Ltd., Cambridge, UK, pp. 321-326.

34. Hatakeyama, T., Hatakeyama H. and Nakamura; K., 1995, Non-freezing water content of mono- and divalent cation salts of polyelectrolytes-water systems studied by DSC, Thermochimica Acta, **253**, 137.

35. Nishinari, K., Food Hydrocolloids in Japan, Food Hydrocolloids, 373-390

36. Nakamura, K. Hatakeyama T., Hatakeyama, H., 2000, TMA measurement of swelling behavior of polysaccharide hydrogels, *Thermochimica Acta,* **252/253**, 171-176.

37. Okuyama, K., Ararnott, S., Moorhouse, R., Walkinshaw, M. D., Atkins, E. D. T. and Wolfullish, CH., 1980, Fiber diffraction studies of bacterial polysaccharides, in *Fiber Diffraction Methods*, French, A. D. and Gardner, K.C. H. eds., ACS Symp. Ser., 141, Am. Chem. Soc., Washinton DC, pp. 411-427.

38. Maret, G., Milas, M. and Rinaudo, M., 1981, Cholesteric order in aqueous solutions of the polysaccharaide xanthan, *Polym. Bulletin,* **4**, 291-297.
39. Ross-Mruphym S. B., Morris, V. J. and Morris, E. R., Molecular viscoelasticity f xanthan polysaccharide, 1983, *Faraday Symp. Chem. Soc.*, 18, 115-129.
40. Tako, M. and Nakamura, S., 1985, Synergistic interaction between xanthan and guar gum, *Carbohydr. Res.*, **138**, 207-313.
41. Quinn, F. X., Hatakeyama, T., Takahashi, M. and Hatakeyama; H. 1994 The effect of annealing on the conformational properties of xanthan hydrogels", *Polymer*, **35**, 1248.
42. Yoshida, T., Takahashi, M., Hatakeyama, T. and Hatakeyama, H. 1995, Annealing induced gelation of xantahn/water systems *Polym.*, **39**, 1119-1122.
43. Takahashi, M. Hatakeyama T. and Hatakeyama, H., 1998, Effect of heat-treatment of solution on the formation of gels of the polysaccharide-water system, *Kobunshi Ronbunsyu,* **55**, 760-767 (Japanese).
44. Takahashi, M. Hatakeyama T. and Hatekeyama, H., 2000, Phenomonological theory describing the behviour of non-freezing water in structure formation process of polysaccharide aqueous solutions, *Carbohydrate Polym.*, **42**, 91-95.
45. Takahashi, M. Hatakeyama T. and Hatakeyama, H., 2001, Temperature and concentration dependency on equilibration of polysaccharide electrolyte hydrosols", in *Recent Advances in Environmentally Compatible Polymers*, (J. F. Kennedy, G. O. Phillips, P. A. Williams and H. Hatakeyama eds.), Woodhead Publishing Ltd., Cambridge, UK, pp. 145-154.
46. Fujiwara, J., Iwanami, T., Takahashi, M., Tanaka, R., Hatakeyama. T. and Hatakeyama H.. 2000, Structural change of xanthan gum association in aqueous solutions, *Termochimica Acta,* **352-353**, 241-246.
47. Iseki, T., Takahashi, M., Hatano, H., Hatakeyama T. and Hatakeyama, H., 2001 "Viscoelastic properties of xanthan gum hydogels annealed in the sol state", Food Hydrocolloids, **15**, 503-506.
48. Harada, T., Misaki T. and Saito H., 1968, Curdlan, a bacterial gel-forming β-1, 3-Glucan, *Achieves of Biochemistry and Biophysics,* **123**, 292-298.
49. Harada, T., 1987, Special bacterial polysaccharides and plysaccarases, *Biochim. Soc. Symp.* **48**, 97-116.
50. Kanzawa, Y., Harada, T., Koreeda, A., Harada, A. and Okuyama K., 1989, Difference of molecular association in two types of curdlan gel, *Carbohydrate Polym.*, **10**. 299-313.
51. Hirashima, M., Takaya, T. and Nishiunar, K., 1997, DSC and rheological studies on aqueous dispersions of curdlan, *Thermichimica Acta*, **306**, 109-114.
52. Tada, T., Tamai, N. Matsumoto T. and Masuda T., 2001, Network structure of curdlan in DMSO and mixture of DMSO and water, *Biopolymer,* **58**, 129-137.
53. Zhang H., Nishinari, K., Williams, M. A. K., Foster, T. J. and Norton, I. T. 2002, A molecular desctiption of the gelation mechanism of curdlan, *Inter, J. Biol Macromol* **30**, 7-16.
54. Hatakeyama, T., Yamauchi, A. and Hatakeyama, H., 1987, Effect of thermal hsteresis on structural change of water restrained in poly(vinyl alcohol) psuedo-gel, *Eur. Polym. J.*, 23, 361-365.
55. Stauffer, S. A. and Peppas, N. A., 1992, Poly(vinyl alcohol) hydrogels prepared by freezing-thawing cyclic processing, *Polym.*, **33**. 3932-3936.
56. Hyon, S-H., Cha, W-I. and Ikada, Y., 1989, Preparation of poly(vinyl alcohol) by low temperature crystallization of the aqueous poly(vinyl alcohol) solution, *Koubnshi Ronbunshuu*, **46**, 673-680.

57. Richardson P. H., Clark A. H., Russell, A. L., Aymard, P. and Norton I. T., 1998, Gelation behaviour of concentrated locust bean gum solutions. *Macromolecules*, **31**, 1519-1527.
58. Tanaka, R., Hatakeyama T. and Hatakeyama, H., 1998, Formation of locust bean gum hydrogel by freezing thawing, *Polymer International*, **45**, 118-126.
59. Kesavan, S. and Prud'homme, R. K., 1992, Rheology of guar and HPG cross-linked by borate, *Macromol.* 25, 2026-2032.
60. Robinson, G, Ross-Murphy, S. B. and Morris, E. R., 1982, Viscosity-molecular weight relationships, intrinsic chain flexibility, and dynamic solution properties of guar galactomannan. *Carbohydrate Res.*, **107**, 17-32.
61. Buren, J. P. V., 1991, Function of pectin in plant tissue structure and firmness, in *Chemistry and Technology of Pectin*, Walter, H. ed. Academic Press Inc., San Diego, 1-22.
62. de Vries, J. A., Hansen, M., Søderberg, J., Glahn,P.-E. and Pedersen, J. K., 1986, Distribution of methoxyl groups in pectin., *Carbohydrate Polym.*, **6**, 165-176.
63. Powell, D. A., Morris, E. R., Gidley, M. J. and Rees, D. A., 1982, Conformations and interactions of pectins II, Influence of residue sequene on chain association in calcium pectate gels, *J. Mol. Biol.*, **155**, 517-531.
64. Iijima, M., Nakamura, K., Hatakeyama, T. and Hatakeyama, H., 2000, Phase transition of pectin with sorbed water", *Carbohydrate Polymers*, **41**, 101-106.
65. Iijima, M., Nakamura, K., Hatakeyama T. and Hatakeyama, H., 2001, Structural Change of water restrained by pectins, in "Recent advances in environmentally compatible polymers", (Kennedy, J. F., Phillips, G. O., Williams, P. A. and Hatakeyama, H., eds.), Woodhead Publishing Ltd., Cambridge, UK, pp. 303-308.
66. Iijima, M., Hatakeyama, T. Takahashi, M. and Hatakeyama, H., 2001 "Thermomechanical analysis of polysaccharide hydrogels in water, *J. Thermal Analysis and Calorimetry*, **64**, 617-627.
67. Hatakeyama, T., Nakamura, K. Yoshida, H. and Hatakeyama, H., 1989, Mesomorphic properties of highly concentrated aAqueous solutions of polyelectrolytes from saccharides", *Food Hydroclloids*, **3**, 301-311.
68. Hatakeyama H. and Hatakeyama, T., 1991, Mesomorphic properties of polyelectrolytes with water, in *Properties of Ionic Polymers Natural and Synthetic*, Salmen L., Htun M. (editors), STFI-meddelande, **a 989**, p123-147.
69. Yoshida, H., Hatakeyama, T. and Hatakeyama, H., 1990 Phase transitions of the water-xanthan system", *Polymer,* **31**, 693.
70. Yoshida, H. Hatakeyama, T. and Hatakeyama H., Effect of water on the main chain motion of polysaccharide hydrogels", Viscoelasticity of Biomaterials, Glasser, W. G. and H. Hatakeyama, H., eds. ACS Symposium, Series **489**, ACS, Washington DC, Ch.14, 218 (1992).
71. Naoi, S., Hatakeyama T. and Hatakeyama, H., 2002, Phase diagram of locust bean gum-, tara gum- and guar gum-water systems, *J. Thermal Anal. Calorimetry*, **70**, 841-851.

Chapter 5

LIGNIN

1.   INTRODUCTION

According to the well-known book Lignins, edited by K. V. Sarkanen and C. H. Ludwig [1], the word lignin is derived from the Latin word *lignum* meaning wood. The amount of lignin in plants varies widely according to the kind of plant. However, in the case of wood, the amount of lignin ranges from ca. 19 - 30 %, and in the case of nonwood fibre, ranges from ca. 8 - 22 %, when the amount is determined according to Klason lignin analysis which is dependent on the hydrolysis and solubilization of the carbohydrate component of the lignified material, leaving lignin as a residue [1-5].

Lignin is usually considered as a polyphenolic material having an amorphous structure, which arises from an enzyme-initiated dehydrogenative polymerization of p-coumaryl (I), coniferyl (II) and sinapyl (III) alcohols (see Figure 5-1).

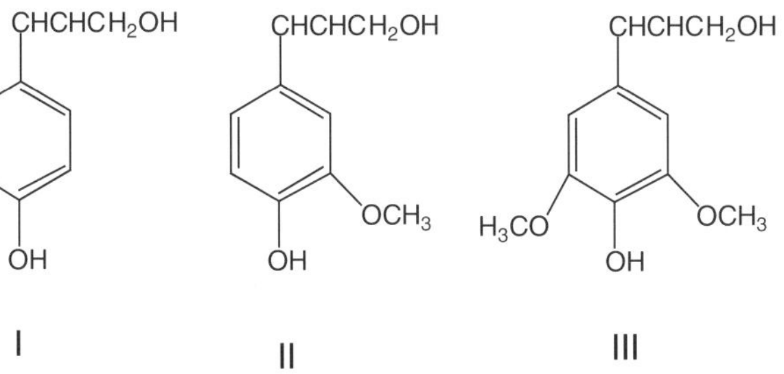

Figure 5-1. Chemical structures of three alcoholic precursors of lignin

The basic lignin structure is classified into only two components; one is the aromatic part and the other is the C3 chain. The only usable reaction site in lignin is the OH group, which is the case for both phenolic and alcoholic hydroxyl groups.

Lignin consists of p-hydroxyphenyl (I), guaiacyl (II), and syringyl (III) structures connected with carbon atoms in phenylpropanoid units, as illustrated in Figure 5-2.

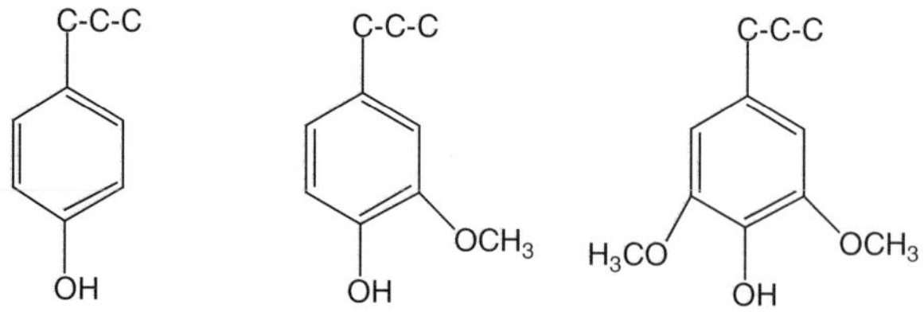

Figure 5-2. Three important structures of lignin. p-hydroxyphenyl (I), guaiacyl (II), and syringyl (III) structures

Lignin is a major component of plants. Scanning electron and ultraviolet micrographs of the cross section of cedar show that lignin is mainly present at inter-cell membranes. An atomic force micrograph of the molecular level structure of lignin is shown in Figures 1-5, 1-6 (Chapter 1).

The structure and amount of lignin in living plants depend not only on plant species but also on location of tissue, age of the plant, and other natural conditions. It is thought that lignin is synthesized enzymatically by the modification of saccharides. Due to the biosynthetic process in living tissues, it is reasonable to consider that lignin has an extremely complex chemical structure as shown in Figure 1-9. At the same time, it is known that lignin exists as a matrix in the architecture of plants, compiled according to the hierarchy of plant organization [5,6].

The above complex constitution evolved in nature, with numerous biocomponents organized and engaged in specific functions, in which lignin works as a matrix component with viscoelastic properties. Lignin having slight hydrophobicity cooperatively affiliates with hydrophilic polysaccharides [4,5].

In this chapter, glass transition behaviour is explained in sections 2 and 3, since the main chain motion is the most important transition behaviour of solid lignin. Local mode relaxation at a low temperature region will be introduced based on viscoelastic and nuclear magnetic resonance spectroscopy (NMR) in section 4. Water-lignin interaction is described in section 5. Thermal decomposition based on thermogravimetry (TG) and

TG-Fourier transformed infrared spectroscopy (FTIR) is considered in section 6. Applications of lignin to various environmentally compatible polymers will be described in Chapters 6, 7 and 8.

2. GLASS TRANSITION OF LIGNIN IN SOLID STATE

Figure 5-3 shows wide angle x-ray (WAX) diffractograms of lignin extracted by various methods. A diffused halo pattern having several peaks indicating a broad structure distribution can be seen. The intra- and intermolecular distance at peak is 0.43 nm and 0.98 nm for dioxane lignin (DL), 0.42 nm and 0.63 nm for milled wood lignin (MWL) [6]. The WAX pattern of atactic-polystyrene (a-PSt) with molecular mass 1.0×10^5 and molecular weight distribution 1.01, a representative synthetic amorphous polymer, is also shown as a reference. Polystyrene shows two distinct peaks at 4.57 nm and 8.8 nm. The former corresponds to average values of intra-molecular distance and the latter corresponds to inter-molecular distance [7]. In contrast, each lignin peak is not as distinct as that of PSt, suggesting that the molecular arrangement of lignin samples has a broad distribution. The WAX patterns shown in Figure 5-3 indicate that lignin is an amorphous polymer having wide distribution of intermolecular distance and lacking any type of higher-order molecular regularity.

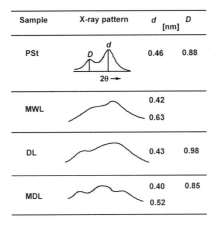

Figure 5-3. Wide angle x-ray (WAX) diffractograms of lignin and polystyrene. PSt: polystyrene (molecular weight 1×10^5, M_w/M_n = 1.01), MWL: milled wood lignin. DL: dioxane lignin, MDL, methylated DL. Measurements; WAX was measured using a Rigaku, 20001 type x-ray diffraction analyzer at 35 kV, 20 A, with a Ni filter using a goniometer. Powder shape lignin was compressed into a pellet and the thickness of the pellet was ca. 1 mm.

Lignin shows no first order thermodynamic phase transitions. Neither thermal nor liquid induced crystallization is known for lignin in the solid state. This indicates that solid lignin takes either the glassy state or rubbery state, depending on temperature, at a temperature lower than thermal decomposition [8]. On this account, local mode relaxation, glass transition and decomposition are expected to be found when lignin is heated from low to high temperature.

2.1 Glass transition of isolated lignin

When a polymer melt is cooled at the isobaric condition, the melt crystallizes at a characteristic temperature (the melting temperature, T_m), if nuclei are formed and the nucleating rate exceeds the cooling rate. However, if the above conditions are not attained, the melt is maintained in a metastable state (super-cooled melt) at a temperature lower than T_m. . On further cooling, the viscosity of super-cooled melt increases and the melt solidifies at a temperature where the configurational entropy of the melt reaches a characteristic value. This glassy solidification temperature is defined as glass transition temperature (T_g). Molecular motion of polymers which is observed such as, specific heat capacity, modules of elasticity, expansion coefficient, dielectric constant, NMR spin-lattice relaxation time, etc. change in a characteristic manner at T_g [3-20]. The glassy state is a thermodynamically non-equilibrium state, and on this account, glass transition behaviour is time dependent and influenced by measurement conditions. In spite of the above facts, the T_g value of each polymer can be observed in a certain definite temperature range, since the change occurs in a drastic manner. Although various experimental techniques, such as viscoelastic measurement, nuclear magnetic resonance spectroscopy, differential scanning calorimetry (DSC) and dielectric measurement are known, DSC is the most widely used in order to measure the glass transition temperature of amorphous polymers [9-18]. Typical DSC heating curves showing glass transition are found in the schematic presentation in Figure 2-5 (Chapter 2). In this book, T_{gi} is used as an index of glass transition temperature.

2.1.1 Glass transition of various kinds of lignin

WAX patterns shown in Figure 5-3 indicate that lignin is an amorphous polymer having wide distribution of inter-molecular distance. Higher order molecular regularity is not observed. Figure 5-4 shows DSC heating curves of different types of lignin, milled wood lignin (MWL), dioxane lignin (DL) and Kraft lignin (KL). As shown in Figure 5-4, baseline deviation due to glass transition is clearly observed. In order to make the thermal history identical, all samples were heated at a temperature 30 K higher than glass transition temperature (T_g) and quickly cooled to room temperature.

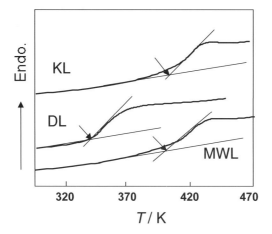

Figure 5-4. DSC heating curves of lignin extracted by various methods. KL (kraft lignin), DL (dioxane lignin), MWL (milled wood lignin). Measurements; Power compensation type DSC (Perkin Elmer), heating rate = 10 K min^{-1}, sample weight = 5 mg, N_2 flow rate = 15 ml min^{-1}.

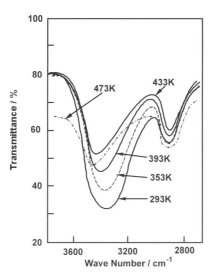

Figure 5-5. Representative IR spectra of milled wood lignin (MWL) at various temperatures. Samples (see footnote of Table 5-1), Powder lignin sample (dried at 10^{-4} mmHg for 48 hrs) was mixed with KBr powder and pressed into a pellet. Measurements; Infrared spectrometer (Perkin Elmer), conformation of the temperature controllable sample holder is shown in Chapter 2, Figure 2.19. Temperature was controlled stepwise.

At around glass transition temperature, intermolecular hydrogen bonding breaks and molecular motion is enhanced [19, 20]. Variation of OH stretching absorption, C=O stretching, aromatic skeletal vibration, C-O stretching and C-O deformation of infrared (IR) spectra of various types of lignin were measured as a function of temperature [21]. Figure 5-5 shows representative IR spectra in a wave number from 2500 to 4000 cm^{-1} of milled wood lignin (Björkman lignin) (MWL) at various temperatures.

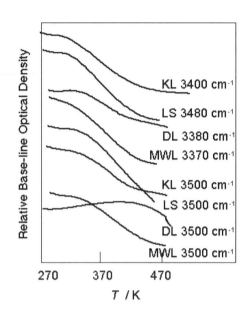

Figure 5-6. Relationship between relative optical baseline density of representative OH stretching band (hydrogen bonded) and temperature of various kinds of lignin. I: KL (3400 cm^{-1}, 3500 cm^{-1}), LS (3380 cm^{-1}, 3500 cm^{-1}), DL (3380 cm^{-1}, 3500 cm^{-1}) and MWL (3370 cm^{-1}, 3500 cm^{-1}). Sample preparation (see Table 5-1 footnote).

The relative baseline optical density was calculated as stated in 2.2.4. Figure 5-6 shows representative curves of relative baseline optical density of OH stretching band of various types of lignin as a function of temperature. The relative optical density of OH stretching band starts to decrease at a temperature lower than glass transition temperature measured by DSC. In contrast, the absorption bands of the aromatic skeletal vibration decreased at a temperature higher than that of OH stretching. Table 5-1 shows the inflection point of relative optical density curves of various types of lignin, together with assignment of each absorption band.

When temperature dependency of WAXS patterns of various types of lignin was measured, a broadening of the halo pattern was observed. Figure

5-7 shows relationship between intermolecular distance (*d*) of DL and temperature. The value of *d* increases at around T_g where endothermic deviation was observed in the DSC curves shown in Figure 5-4.

Table 5-1. Temperature of inflection point of relative optical density curves of various kinds of lignin

Wavenumber / cm⁻¹	Assignment	MWL[*1]	DL[*2]	LS[*3]	KL[*4]
3500		350		330	350
3370-3400	OH stretching	370	390	410	410
1590-1595	Aromatic skeletal vibration	390	410		450
1420-1440	Aromatic skeletal vibration	390	380	459	450
1255-1265	C-O stretching, aromatic (methoxyl)	350	350	350	359
1205-1215	C-O stretching, aromatic (phenol)	370	370	370	370
1025-1035	C-O deformation (primary hydroxyl and methoxyl)	390	350	380	390

*1 MWL; MWL was prepared according to Bjorkman's procedure from spruce (Picea Jezoensis). Purification was carried out by repeated precipitation of dichloroethane-ethanol solution of MWL into ethyl ether.

*2 DL; Dioxane lignin was prepared according to Junker's procedure from Japanese cypress (Cupressauceae obutusa). Purification was carried out by repeated precipitation of dioxane solution of MWL into ethyl ether

*3 LS; Commercially obtained calcium lignosulfonate was purified by gel chromatography.

*4 KL: Commercially obtained Kraft lignin from softwood was purified by repeated precipitation of dioxane solution of MWL into ethyl ether

*5 The absorption may be affected by sulfonate groups.

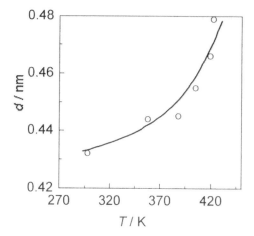

Figure 5-7. Relationship between intermolecular distance (*d*) and temperature of DL. Measurements; x-ray diffractometer (Rigaku Denki), DL powder was filled in a hole with diameter 5 mm in a metal plate with thickness 0.7 mm. Both sides of the hole were

windowed using mica sealed with epoxy resin. This metal plate was set in a sample holder whose temperature was controlled using a temperature controller.

Although T_g values determined by DSC, IR spectrometry and WAXS do not accord well with each other due to the difference in sample preparation and temperature control mode, it is clear that inter-molecular hydrogen bonding breaks in the initial stage (Figure 5-6) and intermolecular distance (Figure 5-7) and heat capacity increase at glass transition.

2.1.2   Effect of molecular mass

Molecular mass and molecular mass distribution (M_w,/ M_n, where M_w is mass average molecular mass and M_n is number average molecular mass) are major factors affecting the molecular mobility of polymers in solid state. Among many kinds of amorphous polymers, glass transition behaviour of polystyrene has been investigated from the view point of molecular weight in the last fifty years.   Since polystyrene is soluble in various organic solvents, a variety of experimental techniques can be applied to measure the molecular weight, such as viscosity measurement, light scattering, sedimentation, gel permeation chromatography, etc.   Furthermore, polystyrene samples having very narrow molecular weight distribution (M_w/ M_n = 1.01 to 1.10) can be synthesized by ionic polymerization.   On this account, it is easy to obtain samples having a wide range of M_w/ M_n.

Glass transition temperature (T_g) of polystyrenes with various molecular weights are reported in a molecular weight ranging from monomer to M_n = 1×10^6. T_g. of polystyrene increases with increasing molecular weight to 5 x 10^4 and is then maintained at a constant value (ca. 360 K). T_g decreases and glass transition temperature range expands with increasing molecular weight distribution. In an oligomeric molecular mass range, the chemical structure of the end group affects T_g [22].   The effect of the end group of poystyrene can be observed in a M_n ranging from monomer to ca. 6-mer [22].

In contrast to polystyrene samples, only a small number of T_g data of lignin having various molecular weights is known [23-25].   Molecular weight and its distribution of lignin markedly depend on isolation conditions. When the lignin samples are examined by analytical methods, such as gel permeation chromatography, it is necessary to solve the samples in organic solvents. In the case of lignin, it is thought that the high molecular weight portion of lignin and/or three-dimensional network portion is not easily soluble and the insoluble portion in organic solvents is excluded by filtration.   It is believed that the molecular weight of purified lignin is considerably lower than the original molecular weight.

Table 5-2 shows M_n and M_w / M_n of KL fractionated by successive precipitation. Molecular weight and molecular weight distribution are measured by gel permeation chromatography (GPC) using tetrahydrofuran (THF) as an eluent. Gel permeation chromatograms of unfractionated and fractionated KL are shown in Figure 5-8. For the M_n and M_w / M_n calculation, polystyrene samples having M_n = 600 to 105 M_w / M_n = 1.1 to 1.01 were used as reference materials. The molecular distribution values of soft wood KL are small in the low molecular weight fractions and are large in the high molecular weight fractions. This indicates that the molecular weight distribution of the original sample is non-Gaussian type distribution. GPC chromatograms also indicate that the low molecular weight portion is distorted in the original sample.

Table 5-2. Molecular mass and molecular mass distribution of unfractionated and fractionated KL

Sample			Symbol	M_w	M_w/M_n
Unfractionated	Hard wood*		I	2600	1.33
Unfractionated			II	4200	1.71
			II-1***		
			II-2	7500	
			II-3	7000	1.54
		KL	II-4	6000	1.48
Fractionated	Softwood**		II-5	4900	1.53
			II-6	4400	1.45
			II-7	4000	1.41
			II-8	4000	1.39
			II-9	3800	1.42
			II-10	2000	1.30

*white fir,**beech

Fractionation was carried out as follows; KL (10g) was dissolved in dioxane and kept in a water bath at 308 K. Water was added dropwise into the above solution with stirring. After keeping the solution at room temperature for 24 hours, the precipitate was separated by centrifugation (5000 rpm). The fraction which had been separated was redissolved in fresh dioxane and reprecipitated with the same ratio of precipitant to solvent as the solution from which it was separated and stored at 308 K for 24 hours then it was centrifuged again.

Glass transition temperature of unfractionated and fractionated lignin was determined by DSC. As shown in Figure 5-9, a baseline gap due to glass transition is observed. It is seen that T_g shifts to the high temperature side with increasing molecular weight. Figure 5-10 shows the relationship between T_g and molecular weight. T_g increases linearly with increasing molecular weight. Ordinarily, T_g levels off when the molecular weight attains a characteristic value. T_g of the fraction having the largest molecular weight does not reach the leveling off point.

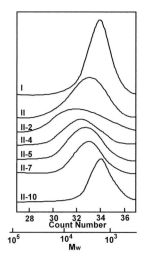

Figure 5-8. Gel permeation chromatograms of unfractionated and fractionated KL. Symbols are the same as Table 5-2.

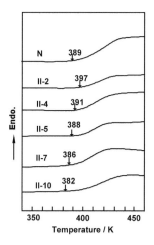

Figure 5-9. DSC heating curves of unfractionated and fractionated lignin samples. Symbols in the figure correspond to those shown in Table 5-2. Measurements; power compensation type DS S (Perkin Elmer), heating rate = 20 K min^{-1}, sample mass = ca 7 mg, N$_2$ flow rate = 10 ml min^{-1}.

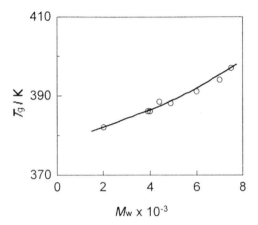

Figure 5-10. Relationship between T$_g$ and molecular mass of fractionated dioxane lignin (DL).

2.1.3 Effect of thermal history

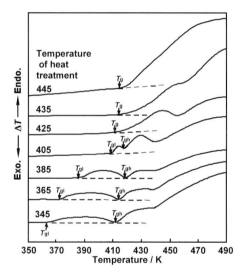

Figure 5-11. DSC heating curves of lignin annealed at various temperatures. Mesurements; power compensation type DSC (Perkin Elmer). heating rate = 20 K min^{-1}, Samples were annealed at a temperature indicated in the figure for 30 min [26].

The glassy state is thermodynamically non-equilibrium. Enthalpy relaxation of amorphous materials will be described in the section 3.2. Since glass transition is a molecular relaxation phenomena, the glass transition behaviour of amorphous polymers, such as lignin, is markedly affected by the thermal history [24]. On this account, when two samples are compared with each other, it is necessary to define the history of the sample in a similar manner.

DSC curves of lignin annealed at various temperatures are shown. As shown in Figure 5-11, sub- T_g is observed by annealing at a temperature lower than T_g. The temperature of sub- T_g shifts to the high temperature side and when the annealing temperature exceeds T_g, sub- T_g disappears. The sub-T_g was found not only in lignin but also in other amorphous polymers such as polystyrene and poly(vinyl chloride). The above facts indicate that the conformation of molecular chains of amorphous polymer has a broad distribution and molecular rearrangement takes place by annealing.

2.1.4   Glass transition of lignin derivatives

Various types of lignin derivatives were prepared such as, acetoxyl lignin, hydroxypropyl lignin and others. The hydroxyl group of lignin is used as a reaction site. Although a large number of attempts have been made in order to prepare new lignin derivatives, T_g values of the samples have not been reported. Ordinarily, molecular motion of lignin is enhanced by the introduction of a large side chain molecule. The glass transition of lignin derivatives shifts to the low temperature side due to the decrease of the amount of hydrogen bonding and expansion of inter-molecular distance. Glass transition of methylated dioxane lignin (MDL) was measured by DSC. Figure 5-12 shows the variation of T_g as a function of methoxyl content. T_g of MDL markedly decreases with methoxyl content.

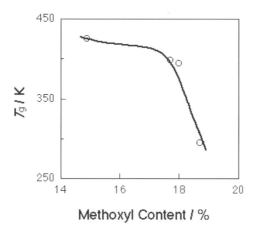

Figure 5-12. . Glass transition temperature of methylated DL with various methoxyl contents. Tg data was obtained by DSC. heating rate = 10 K min^{-1}, sample weight = ca. 8 mg, N$_2$ flowing atmosphere.

T_g decreases by the introduction of acetoxyl group. By the acetylation, it was difficult to convert all hydroxyl groups into acetoxyl groups. On this account, the relative optical density of absorption bands at 1740 cm^{-1} and 1760 cm^{-1} (C=O stretching vibration due to the acetoxyl group) was chosen as a criteria of acethyl conversion rate. In the above calculation, the absorption band at 1600 cm^{-1} due to the aromatic ring vibration was used as an internal standard. Relative values of acetylation were calculated using an index where the IR absorption band reaching the saturated point was assumed.

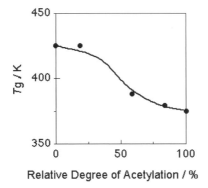

Figure 5-13. Relationship between T_g and relative degree of acetylation of DL

Figure 5-13 shows the relationship between T_g and acetylation rate of acetyl lignin (ADL) [27]. T_g linearly decreases with increasing number of acetyl group. This shows that the molecular mobility increases with the

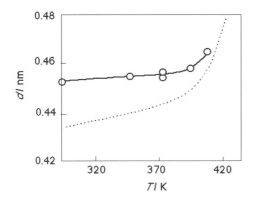

Figure 5-14. Relationships between intermolecular distance (d) of MDL, DL and temperature. Dotted line: DL. Measurements; see Figure 5-7 caption.

introduction of bulky side chains and inter-molecular hydrogen bonding of lignin is disrupted.

Figure 5-14 shows temperature dependence of inter-molecular distance (*d*) of MDL measured by wide angle x-ray diffractometry (WAX). The variation of *d* of DL is also shown as a dotted line. d increases with the introduction of acetyl group and, at the same time, molecular expansion starts at a temperature lower than that of DL. Both x-ray and DSC results indicated that the introduction of bulky side chains expand the molecular distance and enhance the molecular motion.

3.   HEAT CAPACITY AND ENTHALPY RELAXATION OF LIGNIN

3.1   Heat capacity of lignin at around glass transition

Specific heat capacities (C_p) at an isobaric condition are generally measured by adiabatic calorimetry or differential scanning calorimetry (DSC). In order to obtain reliable C_p data by adiabatic calorimetry, a complex home-made apparatus, trained experimental personnel and a large amount of pure samples are necessary. In contrast, C_p is more easily obtained by DSC and C_p data well accord with those obtained by adiabatic calorimetry, although the precision of the data is low [28]. C_p values of dioxane lignin measured by DSC are shown in Figure 5-15 [29]. C_p variation due to glass transition is clearly seen at around 400 K. C_p values are almost the same as other amorphous polymers such as polystyrene.

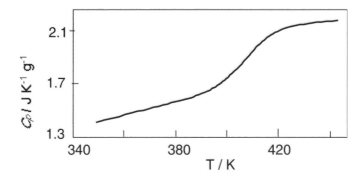

Figure 5-15. Specific heat capacity of dioxane lignin   (DL). Measurements; Powder compensation type DSC (Perkin Elmer), heating rate = 10 K min^{-1}, sample mass = ca. 5 mg, reference materials, sapphire, N$_2$ gas flow rate = 30 ml min^{-1}.

3.2   Enthalpy relaxation of lignin

Polymers in the glassy state below the glass transition temperature are not in thermodynamic equilibrium and relax towards equilibrium with time. Relaxation time distribution necessarily exists in the glassy state, since molecular chains are unable to solidify homogeneously and simultaneously at certain time intervals. As described in section 1, the chemical structure of lignin is complex. On this account, it is considered that molecular chains of lignin cease their molecular motion in unequlibriated conditions. The rate of solidification affects the enthalpy level of glassy lignin. When a lignin sample is slowly cooled from the melt, enthalpy of the glassy state decreases. In contrast, the enthalpy increases when the sample is quenched, since molecular chains are more randomly frozen than those of slowly glassified samples.

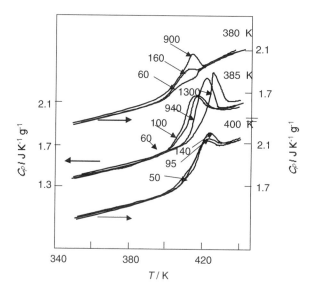

Figure 5-16. Heat capacity of dioxane lignin annealed at 380, 385 and 400 K for various times. Annealing time is shown in the figure (min). Measurements; Power compensation type DSC (Perkin Elmer), heating rate = 10 k min^{-1}, sample mass = ca. 7 mg. The samples were once heated to 460 K, quenched to 320 K and then C_p was measured.

For this reason, the experimentally measured enthalpy of glassy polymers decreases as a function of time if the sample is maintained below T_g. This phenomenon is called enthalpy relaxation and is monitored through heat capacity change at glass transition. In the presence of enthalpy relaxation, the mechanical, transport and other physical properties of the polymer vary as a function of temperature and time [30, 31]. Gas diffusion through polymer membranes can decrease by as much as two orders of magnitude due to enthalpy relaxation at ambient temperature. The stress-strain curves of glassy polymers reveal more brittle behaviour as enthalpy relaxation proceeds.

Instead of slow cooling, similar relaxation behaviour is observed by annealing the sample at a temperature lower than T_g. The quenched sample whose DSC curve was presented in Figure 5-15, was annealed at at 380, 385 and 400 K for various times. DSC curves of annealed samples are shown in Figure 5-16 [29].

Precise analysis of enthalpy relaxation is not possible due to the non-equilibrium nature of glassy polymers above and below the glass transition. Enthalpy relaxation can be characterized under certain limiting assumptions. The procedure is found elsewhere [11, 15].

3.3   Glass transition of poly(hydroxystyrenes) related to lignin

Polystyrene derivatives having three major structures of lignin (see Figure 5.3) as a pendent were synthesized as model polymers of lignin [32-39].   They are poly(4-hydroxystyrene) (PHS), poly(4-hydroxy, 3-methoxystyrene) (PHMS) and poly(4-hydroxy, 3,5-methoxystyrene) (PHDMS) Chemical structures of the above polymers are shown in Figure 5-17. Molecular mass ranged from 1.0 x 10^4 to 1.0 x 10^5 and molecular mass distribution (= M_w/M_n) calculated by gel permeation chromatography ranged from ca. 3.0 to 4.0.

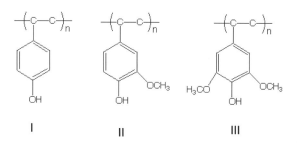

Figure 5-17. Chemical structure of polystyrene and its derivatives related to lignin. PSt: polystyrene, PHS: poly(4-hydroxystyrene), PHMS: poly(4-hydroxy, 3-methoxystyrene). PDMS:poly(4-hydroxy, 3,5-methoxystyrene). Sample preparation; synthesis.

Figure 5-18 shows DSC heating curves of polystyrene (PSt) and three kinds of polystyrene derivatives, PHS, PHMS and PHDMS.. Each sample was heated to a temperature 40 K higher than glass transition temperature (T_g) and quenched to ca. 298 K and then heated at 10 K min^{-1}. T_g varied in

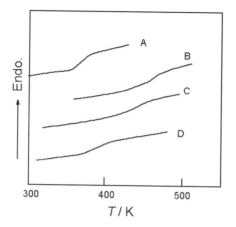

Figure 5-18. DSC heating curves of polystyrene and its derivatives related to lignin. .A: polystyrene, B poly(4-hydroxystyrene), PHMS: poly(4-hydroxy, 3-methoxystyrene). PDMS: poly(4-hydroxy, 3,5-methoxystyrene); Measurements; heat-flux type DSC (Du Pont), sample mass = ca. 5 mg, heating rate = 10 K min^{-1}.

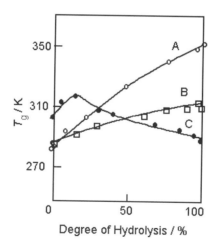

Figure 5-19. Relationships between T_g and degree of hydrolysis of polyhydroxystyrene derivatives, A: poly(4-hydroxystyrene), B: poly(4-hydroxy, 3-methoxystyrene). PDMS: poly(4-hydroxy, 3,5-methoxystyrene).

the order of PHS (455 K) > PHMS (415 K) > PHDMS (381 K) >PSt (366 K). By introduction of the hydroxyl group to 4-position of the aromatic group, T_g markedly increases.

By introduction of the methoxyl group to 3-position of the aromatic group, T_g decreases, since the bulky side group expands intermolecular distance. It was confirmed that biodegradability is enhanced by the introduction of the methoxy group to the 40-poistion [40]. When two methoxy groups are attached to 3- and 5-position, the effect of the hydroxyl group diminishes, since hydrogen bond formation is disturbed geometrically. It was also found that the position of hydroxyl group [39] and number of hydroxyl group [35, 36] affect the T_g value of PHS

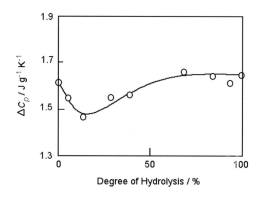

Figure 5-20. Relationship between T_g and heat capacity difference at T_g (ΔC_p) of poly(4-hydroxystyrene) at 373 K.

4.   MOLECULAR RELAXATION

4.1   Viscoelastic measurements

Isolated lignin lacks flowability at a temperature higher than T_g. It is difficult to make strong film either by solvent casting or hot press methods. When the mechanical and viscoelastic properties of polymers are measured, it is crucial to measure the size of the sample precisely in order to obtain quantitative data. Torsional braid analysis (TBA) is an experimental technique using an inert support to measure the viscoelastic behaviour. By TBA, the relative value of rigidity and retardation can be measured as a function of time, although dynamic modulus cannot be obtained quantitatively. This is not only due to the fact that the size of the sample is not defined, but also because the frequency depends on temperature.

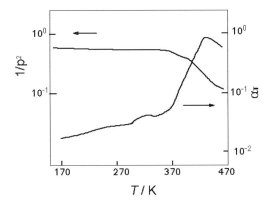

Figure 5-21. Logarithmic decrement (α_T) and relative rigidity (p^{-2}) of dioxane lignin (DL). Preparation of DL (see Table 5-1 footnote), Methoxyl content = 15.4 %. After fixing the braid with lignin to the sample holder, the braid was dried at 293 K for 48 hours under the vacuum 10^{-2} mmHg. Immediately before the measurement, the sample was annealed at 373 K for 2 hrs. Measurement conditions; temperature range 150 to 460 K, heating rate = 1 K min^{-1}, torsion angle 5 degrees [41].

Ordinarily, glass fibre braid is used as a support for TBA. Lignin powder is solved in an organic solvent and the glass support is dipped in the solvent. Then the support with lignin is dried in vacuo in order to eliminate the organic solvent, since a trace amount of solvent works as a plasticiser and molecular motion is easily enhanced at a temperature lower than T_g of dry

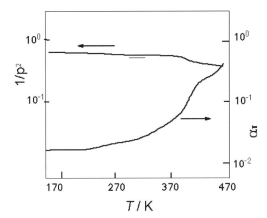

Figure 5-22. Logarithmic decrement (α_T) and relative rigidity (p^{-2}) of Kraft lignin (KL). Preparation of KL (see the footnote of Table 5-1) Methoxyl content = 14.4 %. Measurement conditions; see Figure 5-21 caption.

sample. Figures 5-21, 5-22 and 5-23 show rigidity and increment of dioxane lignin (DL), kraft lignin (KL) and lignosulfonate (LS). The rigidity of DL and KL decreases at around 390 K in the decrement curve. LS shows no significant changes due to strong ionic bondings. Local mode relaxation is not clear except for DL. As shown in the increment curve of Figure 5.19, a shoulder peak is observed at around 330K. The local mode relaxation of lignin will be described in section 3.2 of this chapter,

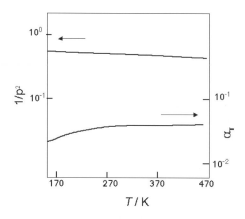

Figure 5-23. Logarithmic decrement (α_T) and relative rigidity (p^{-2}) of calcium lignosulfonate (CaLS). Preparation of LS; see Table 5-1 footnote. Methoxyl content = 12.1 %. Measurement conditions; see Figure 5-21 caption.

In order to measure the dynamic viscoelastic property of lignin, other mechanically inert samples, such as cellulose were used. Filter paper was used as a suitable support for lignin, since significant change in the mechanical properties is not observed at a temperature higher than room temperature to 470 K. At the same time, lignin shows good affinity with cellulose. When the dynamic modulus of DL supported by filter paper is measured, tan δ of DL showed a broad peak at 433K with a shoulder peak at 393 K. In the case of KL, tan δ peak was observed at 408 K.

The mechanical properties of lignin supported by filter paper were measured as a function of temperature as shown in Figure 5-24 The strength of KL decreased as the temperature increased and decreased markedly near the glass transition temperature at around 400 K. The decrease in the strength of DL was observed at 343 and 400 K. In elongation *versus* temperature curves, two peaks are observed for DL and one peak for KL. It is clear that temperature dependence of tensile properties accords well with viscoleastic properties.

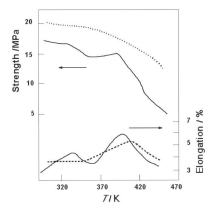

Figure 5-24. The temperature dependence of the tensile properties of lignin supported by filter (cellulose) paper. Solid line: DL, Dotted line: KL. Samples; Dioxane lignin (DL) and Kraft lignin (KL) (preparation, see Table 5-1 footnote). Filter paper was immersed in 30 % tetrahydrofuran solution of lignin, and dried at room temperature for 3 days in vacuo 10^{-4} mmHg at 323 K for 3 days. The lignin content was ca. 40 % in mass. Measurements; Testing conditions, Instron type mechanical tester (Shimadzu), sample length= 50 mm, crosshead speed =10 mm min^{-1}, temperature increased stepwise [42].

4.2 Broad-line nuclear magnetic resonance spectroscopy

Molecular motion of lignin in solid state is measured by broad line nuclear magnetic resonance spectroscopy (b-NMR) as a function of temperature in a wide temperature range. Using b-NMR, not only the main chain motion which has already been described in the former sections, but also local mode relaxation of lignin can be measured [20, 43].

4.2.1 Line shapes and line-width changes

Figure 5-25 shows half of the first derivatives of the proton absorption of dioxane lignin (DL) as a function of the applied magnetic field. As shown in this figure, a broad peak is observed at 120 K, suggesting that the molecular motion appears as one component. At 290 K, a narrow component together with a broad one is observed indicating that a part of the lignin molecules starts to be mobile at this temperature. At 460 K, both components become sharper and narrower. Similar first derivative curves are obtained for Kraft lignin (KL) and lignosulfonate (LS).

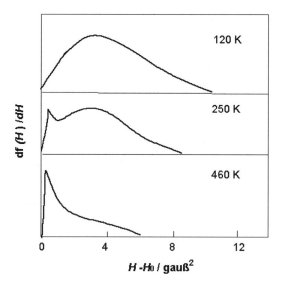

Figure 5-25. First derivative curves of the proton resonance as a function of the applied magnetic field of dioxane lignin.. Apparatus; broad-line NMR (JEOL); . sweeping magnetic field method. Temperature range; 100 to 470 K, temperature increased stepwise, accuracy +/- 0.1 K.

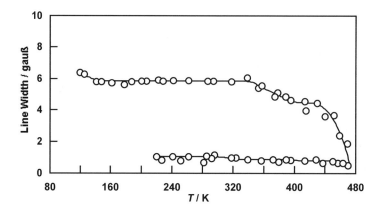

Figure 5-26. Variation of line width of narrow and broad components with temperature of dioxane lignin.

The line width (gauss) of each component was determined from the distance in gauss between the corresponding maximum and minimum of the first derivative curves at each temperature. Figures 5-26, 5-27. and 5-28 show the relationship between line width and temperature of three kinds of lignin. It is clearly seen that the narrow component appears at around 240 K for three kinds of lignin.

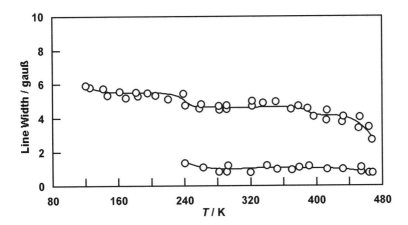

Figure 5-27. Variation of line width of narrow and broad components with temperature of Kraft lignin.

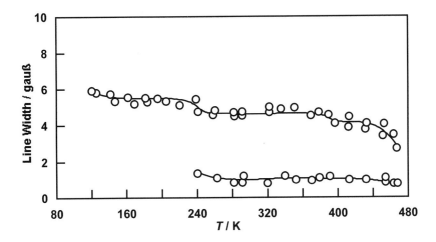

Figure 5-28. Variation of line width of narrow and broad components with temperature of lignosulfonate.

In general, magnetic dipolar broadened resonance starts to become narrow, when the atomic configuration concerning the observed nuclei significantly changes with a frequency which is coincident with the order of the frequency width of the broadened line. At low temperature, quantitative assignment of the line-width change is possible using several approximate relationships, if rigid structure line width, frequency width of resonance line, line-shape parameters etc. are known. In the case of lignin, the low temperature line-width change was assigned as the rotation of methyl groups referring to the results obtained for simple low molecular mass compounds [1,2]. Calculated activation energy estimated for the motion of methyl groups in lignin is shown in Table 5-3.

Table 5-3. Activation energy estimated from the motion of methyl groups in lignins

Lignin sample	Activation Energy / kJ
Dioxane lignin	1.1
Kraft lignin	1.1
Lingosufonate	1.3

4.2.2   Main chain motion and local mode relaxation

The second moment was calculated using differential curves at each temperature and the results are shown in Figures 5-29, 5-30 and 5-31. The second moment decreases discontinuously at a temperature summarized in Tables 5.4 and 5.5 for three kinds of lignin.

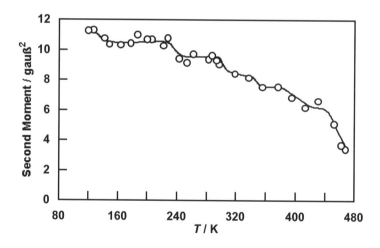

Figure 5-29. Relationship between second moment of dioxane lignin and temperature.

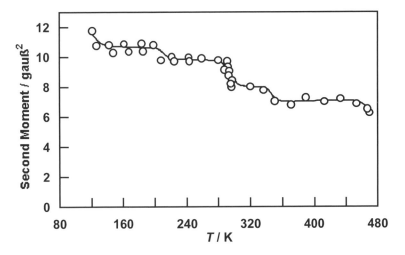

Figure 5-30. Relationship between second moment of Kraft lignin and temperature.

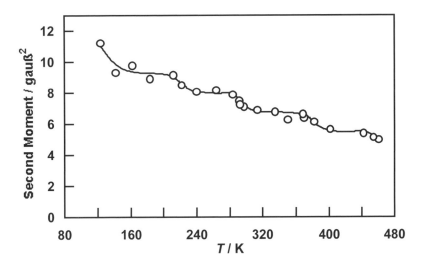

Figure 5-31. Relationship between second moment of lignosulfonate and temperature.

The molecular relaxation observed is designated α, β, γ and δ from the high to low temperature. As shown in Table 5.4, three local mode relaxations are recognized. In order to identify the type of molecular motion of each relaxation by semi-quantitative calculation, information on the chemical structure of lignin is necessary. When empirical formulas of the isolated lignin were assumed to be a single phenylpropane monomer, the second moment can be calculated, i.e. in the softwood milled wood lignin, each monomer unit contains about 1.0 methylene group, about 1.8 methine groups, about 1.2 total amount of hydroxyl groups and about 2.7 aromatic proton. Calculated and observed results are summarized in Table 5-6

Table 5-4. Temperature region of local mode relaxation of various kinds of lignin*

Lignin Sample	Temperature / K			
	δ	γ	β $\beta1$	$\beta2$
Dioxane lignin	110-150	220-250	290-310	330-380
Kraft lignin	110-130	200-220	290-300	330-360
Lingnosufonate	110-150	210-230	280-310	360-400

Details of the samples are shown in Table 5-1 footnote

Table 5-5. Main chain motion (glass transition) of various kinds of lignin measured by b-NMR, DSC and viscoelastic measurement (TBA)

Lignin sample	T_g / K		
	b-NMR	DSC*1	TBA*2
Dioxane lignin	430	415	438
Kraft lignin	450	455	433
Lingnosufonate	450	465	NF
Milled wood lignin	NM	41	NM

*1 T_{gi}; see Figure 2-8.

*2   Peak temperature of decrement (Δ)

NF: not found, NM: not measured

Table 5-6. Observed and calculated second moment (gauss^2) of various kinds of lignin

	DL		KL		LS	
	Observed*	Calculated**	Observed	Calculated	Observed	Calculated
Rigid state	11.3	11.7	11.8	12.6	13.2	12.2
Hindered motion of methyl group	10.6	10.5	10.7	11.4	11.5	10.9
Hindered motion of side chain	9.7	10.0	9.8	10.5	10.1	10.1

* observed values

** calculated values

The local mode relaxations of lignins are evaluated based on the results obtained by b-NMR together with those by viscoelastic measurements and DSC, using not only lignins but also low molecular weight model compounds and polymers having lignin related structure in the main or side chains. The variation in nuclear magnetic resonance with temperature for the lignins shows δ, γ and β relaxations, i.e. (1) δ relaxation is attributed to the hindered rotation of methyl groups which are mainly present as methoxyl group, (2) γ relaxation is attributed to the hindered motion of the side chains of the lignins and (3) β relaxation concerns the hydroxyl groups, the low temperature side is due to hindered motion of the hydroxyl groups in loosely packed molecular chains and the high temperature side is due to the hydroxyl groups which form stable intermolecular hydrogen bondings.

4.3   Molecular motion of lignin measured by spin probe method

It is known that stable radical species exist in lignin in the solid state [44]. Ordinarily, radicals are formed in synthetic polymers by irradiation, such as UV or γ-ray irradiation. When the polymers with stable radicals are heated, the number of spins or radical species varies due to change in environment. On this account, by electron spin resonance spectroscopy, molecular motion of polymers can be measured as a function of temperature. In the case of lignin, irradiation by ionizing sources is not needed, since isolated lignin has radicals without any treatment. Lignin radical scarcely decays regardless of lignin species when it is maintained at room temperature for a long time. By increasing temperature, ESR spectra varied as shown in Figure 5-32.

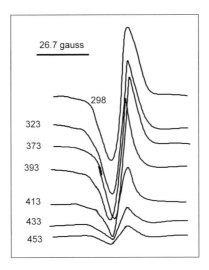

Figure 5-32. ESR spectra of radicals trapped in milled wood lignin (MWL) at various temperatures. (ESR spectra are compared at the same gain). Numerals in the figure show temperature. Measurement conditions; ESR (JEOL). Temperature range 290 to 470 K, temperature increased stepwise.

Number of radicals was measured using reference material (diphenylpicrylhydrazyl, DPPH) whose number of radicals is known. As shown in Figure 5-33, the number of radicals decreases at around 390 K which corresponds to the temperature range of β relaxation measured by b-NMR. By breaking the inter-molecular hydrogen bonding, lignin main chains are gradually enhanced and radicals trapped in the stable molecular chains disappear.

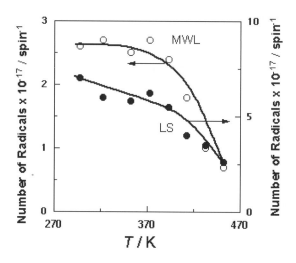

Figure 5-33. Relationship between number of radicals (spins / g) and temperature of lignosulfonate (LS) and milled wood lignin (MWL). ESR spectrometer with double cavity was used for quantitative measurement of spins. DPPH was used as standard material.

5.   LIGNIN-WATER INTERACTION

Biopolymers, such as lignin and polysaccharides, are synthesized in a living organ in the presence of an excess amount of water. The structure and function of lignin are necessarily related with the structure and function of water molecules. When water molecules are strongly restrained by the hydrophilic group of lignin, water molecules are separated from the water cluster and behave as a part of lignin molecules. In this section, glass transition behaviour of lignin with various amounts of water is described. At the same time, structural change of water restrained by lignin is also described

5.1   Water sorption of lignin

Isothermal water absorption measurement is an established experimental technique to characterize the water sorption properties of hydrophilic materials including polymers.   A well dried sample is placed in an atmosphere controlled at a constant relative humidity (RH) at a constant temperature. Samples are maintained long enough to attain the constant mass.   An atmosphere controlled at a constant RH can be prepared using saturated aqueous solution of various inorganic materials. The experimental technique is shown in section 2.2 in chapter 2

When lignosulfonate is stored at various RH's, with increasing stored time, the amount of adsorbed water increases and reaches a constant value. A typical case II type sorption isotherm is obtained as shown in Figure 5.32. From the comparison between the obtained absorption pattern shown in Fig. 5.32 and Fig. 2-x, the absorption curve of DL is classified into Langmuir type.

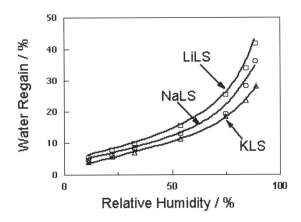

Figure 5-34. Sorption isotherm of LS with different counter ions at 298 K.

Table 5-7. The BET constants obtained and the hydrophilic group content (sulfonate and phenolic hydroxyls) of the lingosulfonate (LS)

Sample	c	m_0 /g g^{-1}	m_0 /meq g^{-1} m_0 / meq g^{-1}	Hydrophylic groups / meq g^{-1}
Li-LS	14,11	7.90×10^{-2}	4.30	3.17
Na-LS	15.60	6.40×10^{-2}	3.56	3.17
K-LS	18.00	5.90×10^{-2}	3.28	3.17

Brunauer-Emmett-Teller (BET) showed that adsorption energy can be calculated from the absorption curves at a low relative humidity range (see section 2.2.2 of Chapter 2) Table 5-7 shows c and m_0 together with hydrophilic group content of lignosulfonates.

5.2   Effect of water on glass transition

Glass transition of lignins in the presence of water was measured by DSC in a similar manner to the cases of polysaccharides. Figure 5.33 shows relationship between T_g and W_c of DL T_g decreases more than 100 K in the presence of a trace amount of water. When W_c exceeds 0.05, T_g maintains an almost constant value. When large substituent groups, such as methoxyl group or acetyl group, are introduced to lignin, T_g in the dry state decreases, since bulky side chain enhances the main chain motion. However, T_g decrease by adding a small amount of water is smaller than that of the original lignin. This is due to the fact that the hydrophilic groups are used as a reaction site and hydrophobicity of molecules increases.

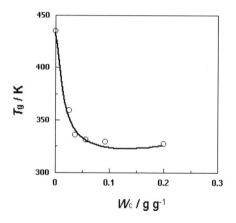

Figure 5-35. Relationship between glass transition temperature (Tg) and water content (Wc) of dioxane lignin (DL). T_g was determined by DSC heating curve. Measurements; power compensation type DSC (Perkin Elmer), heating rate = 10 K min⁻¹, sample mass = ca. 5 mg, aluminium sealed type pan.

By introduction of substituent groups, the effect of water on glass transition behaviour varies. As shown in Figure 5-35, Tg value levels off at a characteristic Wc. The Wc depends on hydrophylicity of lignin. Figure 5-36 shows the relationship between the characteristic W_c^* where T_g values reach a constant value and relative degree of acetylation. With increasing relative degree of acetylation, W_c* decreases, suggesting that hydrophobility of lignin increases.

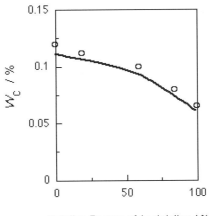

Figure 5-36. Relationship between the characteristic $W_c *$ and relative degree of acetylation. $W_c *$: W_c where T_g value levels off.

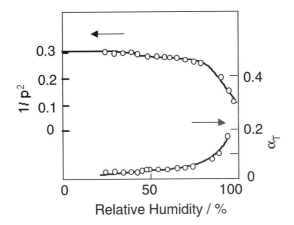

Figure 5-37. Relative rigidity, p^{-2} (p is the period of the oscillation) and decrement (α_T) of NaLS as a function of relative humidity. Measurements; Torsion braid analyzer (self made apparatus), glass fibre was used as a support, temperature = 298 K, torsion angle = 5 degrees.

Molecular enhancement of lignin was also confirmed by viscoelastic measurements in humid conditions. Using a humidity controller, torsion braid analysis (TBA) was carried out at various temperatures as a function of humidity [45 ,46]. Figure 5.35 shows relative rigidity p^{-2} (p is the period of the oscillation) and decrement (α_T) of sodium lignosulfonate (NaLS) as a function of relative humidity. Glass braid was dipped in lignin solution and dried. The braid was twisted and decay of oscillation was measured as a function of time. By TBA, frequency depends on temperature (or in this case humidity) and size of sample cannot be determined. The value p-2 corresponds to dynamic Young's modulus (E') and α_T relates with tan δ measured by stress oscillation method, however, the values are not absolute. As shown in Figure 5-37, p^{-2} decrease at around RH = 80 %. When temperature increased, the temperature where p^{-2} starts to decrease shifted to the low RH side.

5.3   Bound water restrained by lignin

Phase transition of water restrained by lignin was measured by DSC in a temperature range from 150 to 320 K. As described in section 3.1 in Chapter 3, the amount of bound water is defined as the summation of non-freezing water and freezing bound water as defined in equation 3.-5 of Chapter 3. In the case of lignin, the freezing bound water was detected at around W_c= 0.1 g g^{-1} and the maximum amount of freezing bound water was 0.05 g g^{-1} as shown in Figure 5-38.

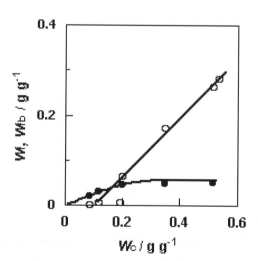

Figure 5-38. Relationship between freezing bound water content (W_{fb}) free water (W_f) and water content (W_c) of dioxane lignin. Open circles: W_{fb}, Filled circles: W_f.

The amount of non-freezing water (W_{nf}) increases with increasing W_c even after the amount of freezing bound bound is saturated. It seems that a characteristic amount of water is required for the homogeneous diffusion of water in lignin molecules. The maximum amount of non-freezing water was ca 0. 12 g g^{-1}, as shown in Figure 5-39.

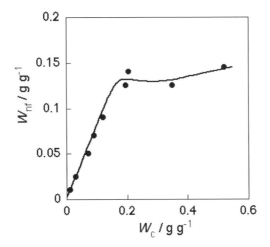

Figure 5-39. Relationship between non-freezing water content (W_{nf}) and water content (W_c) of dioxane lignin.

5.4 ^{1}H and ^{23}Na nuclear magnetic relaxation

Sodium lignosulfonate (NaLS) is a polyelectrolyte and is a water soluble polymer. Due to the presence of sodium ions, it is possible to measure relaxation times of both ^{1}H and ^{23}Na. ^{1}H and ^{23}Na nuclear magnetic relaxation (NMR) times of sodium lignosulfonate (NaLS)-water systems are measured in a water content ranged from 0 to ca. 3.0 g g^{-1} and in a temperature range from 200 to 300 K. NMR relaxation times were measured by decreasing temperature stepwise from 300 to 200 K. In order to coordinate the results obtained by NMR and those of DSC, the phase transition of sorbed water on lignin was measured by cooling. Figure 5-40 shows DSC cooling curves of NaLS-water systems with various water contents. When W_c is 0.46 g g^{-1}, a broad exothermic peak is observed. Two exothermic peaks are observed when W_c exceeds 0.71 g g^{-1}. The high temperature peak shifts to the low temperature and enthalpy of transition decreases with increasing W_c. In contrast, a large crystallization peak shifts to the high temperature side and enthalpy of the peak increases with increasing W_c.

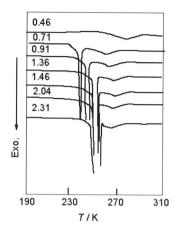

Figure 5-40. DSC cooling curves of water restrained by lignosulfonate. Measurements; power compensation type DSC, cooling rate = 10 K min^{-1}, sample mass = ca. 3 mg, sealed type aluminum pan.

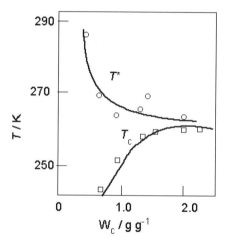

Figure 5-41. Phase diagram of water-NaLS systems established using DSC cooling curves.

Using peak temperatures obtained by DSC cooling curves, a phase transition diagram was established as shown in Figure 5-41. In the figure, the peak temperatures are defined as T^* and T_c. T_c attributed to the crystallization temperature of water in the system and T^* is to transition temperature from liquid to mesophase. When NaLS-water system with a W_c less than 1 g g^{-1} is observed by polarizing microscopy, colour change was detected, although the pattern was not clear like the liquid crystals observed for polysaccharide-water systems. On this account, it is thought that a part of the molecular chains of NaLS is aligned, although all of the molecular chains are not arranged in a fully extended structure.

The longitudinal relaxation times (T_1, sec) of ^{1}H NMR of water in the NaLS-water systems are shown in Figure 5-42 as a function of inverse absolute temperature (K^{-1}). The T_1 value decreases in a temperature higher than 250 K where a minimum value is observed. When temperature exceeds the minimum value, T_1 increased with decreasing temperature. Water molecules in the system become rigid regardless of water content. Figure 5-43 shows the relationship between T_2 value for ^{1}H of water and the inverse absolute temperature. The T_2 values represent the average motion of water molecules in the system. Therefore, it is clear that the molecular motion is more restricted at a lower temperature and more mobile in the higher temperatures [47].

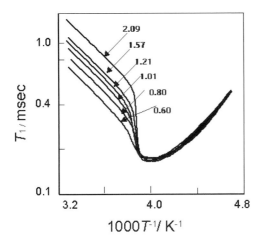

Figure 5-42. Relationship between ^{1}H longitudinal relaxation time of water in the water NaLS systems. Numerals in the figure show water content g g^{-1}. T_1 was measured by 180-τ-90 degree pulse method.

It is only possible to distinguish a limited number of perturbed sites by NMR. On this account, in NaLS-water system, if two sites represent free water fraction and bound water fraction then the equations from 3.13 to 3.20 derived in Chapter 3 can be applied.

The equation 3.19 is a function of $\omega_0\tau_c$. When corelation time (τ_c) values calculated from the equation 3.19 of Chapter 3 are plotted against the inverse absolute temperature, the τ_c values depend on temperature not on water content. The values increased from 3 x 10^{-8} sec at 260 K to ca. 3 x 10^{-7} sec at 200K. The above fact indicates that the bound water in the NaLS-water system is in the state between viscous liquid and non-rigid solid in this temperature range. When the ln τ_c is plotted to the inverse absolute temperature, a linear relationship can be obtained. The temperature dependency of τ_c is expressed by the Arrehnius equation as shown in the equation 3.20. E_a, the activation energy for the relaxation process of the bound water was ca. 24 KJ mol^{-1}.

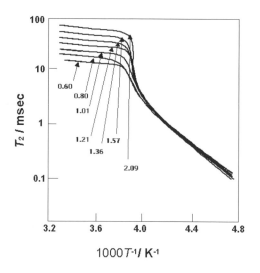

Figure 5-43. Relationship between ^{1}H transverse relaxation time (T_2) of the water NaLS systems. T_2 was measured by the Meiboom-Gill variant of the Carr-Purcell method. In the case of very rapid relaxation, the free induction decay method was used.

Figure 5-44 shows T_1 values of ^{23}Na in the NaLS-water systems as a function of inverse absolute temperature. At a temperature lower than 253 K, T_1 could not be measured by 180-τ-90 degree pulse method due to the extreme broadening of the line width of the peak. The relaxation rate (ln T_1^{-1}) was plotted against the inverse absolute temperature and the apparent activation energy (E_a) was obtained. The calculated value was 12 kJ mol^{-1} which accorded well with E_a of ^{23}Na in persulfonate ionomers with water.

Figure 5-45 shows the relationships between T_2 values of ^{23}Na and inverse absolute temperature. Two types of transverse relaxation, slow and fast relaxation, can be seen. This indicates that the transverse relaxation decays in a non-exponential manner due to quadrapole relaxation. The transverse relaxation brought about quadrapole interaction and is thought to be the summation of two or more decaying exponentials. As shown in Figure 5.43, the long transverse relaxation time (T_{2s}) decreases suddenly at around 258 K, in contrast, the short T_2 (T_{2f}) decreases gradually without any discontinuous change. The sudden decrease of T_{2s} is attributable to the crystallization of water by which mobility of ^{23}Na is restricted. It is thought that the motion of freezing water affects the Na ion attaching to LS in an indirect manner.

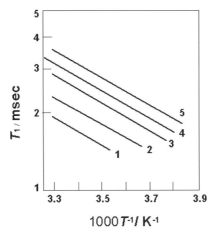

Figure 5-44. Relationship between ^{23}Na longitudinal relaxation time (T_1) of in the NaLS systems as a function of inverse temperature. Water content 1: 0.62, 2: 0.76, 3: 1.02, 4: 1.15, 5: 1.38 g g^{-1}.

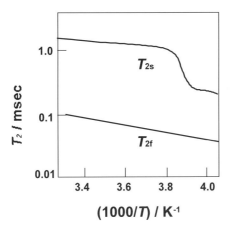

Figure 5-45. Relationship between ^{23}Na transverse relaxation time (T_2) of in the NaLS systems. T_{2s}: T_{2slow}, T_{2f}: T_{2fast}. W_c ranges from 0.6 to 1.38 g g^{-1}.

6.   THERMAL DECOMPOSITION

6.1   TG and TG-FTIR

Considerable number of studies on thermal decomposition of lignin has been reported [2, 48-52], since it is important to obtain gases evolved from lignin in order to use lignin as a raw material for chemical reaction. Thermogravimetry (TG) has extensively been used in order to investigate thermal decomposition behaviour. Figure 5-46 shows TG curves of solvolysis lignin which is obtained as a by-product in organosolve pulping of wood with aqueous cresol without an acid catalyst [53]. As shown in Figure 5.44, mass decrease occurs in two steps at around 570 K and 670K. TG derivatogram also shows a shoulder peak in the low temperature side of the main peak. Starting temperature of decomposition is observed at around 470K and peak temperature is ca. 640 K. TG curves of other types of lignin are found elsewhere [1, 7- 9] It is thought that the first step decomposition is attributable to dehydration from the hydroxyl group located to the benzyl group and this first step decomposition was not observed by acetylation. The second step decomposition is due to heterolysis and homolysis of β-aryl-ether linkage. It is thought that C(aryl)-C(alkyl) bondage breaks at a temperature higher than 570K.

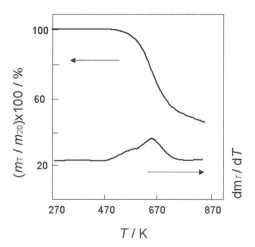

Figure 5-46. TG curve and TG derivatogram of solvolysis lignin. Sample; by-product in organosolve pulping of Japanese beech (*Fagus crenata*) resolved in cresol at 458 K without an acid catalyst. Sample preparation and purification is found elsewhere [53]. Meaurements; Heat-flux type DSC (Seiko Instruments), heating rate = 10 K min^{-1}, N_2 flow rate = 30 ml min^{-1}.

As described in section.1.1of Chapter 2, by simultaneous measurements using TG-FTIR, not only decomposition temperature and mass residue but also gas evolved during decomposition can be identified. Figure 5-47 shows a three dimensional diagram of FTIR spectra among absorbance, wave number and temperature of Kraft lignin [54]. Variation of characteristic absorption bands is shown in Figure 5-48

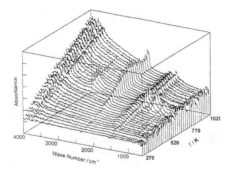

Figure 5-47. TGFTIR of Kraft lignin. Measurements; TG (Seiko Instruments) –FTIR (Jasco FTIR 7000 spectrometer), heating rate = 20 K min⁻¹, N₂ flow rate = 100 ml min⁻¹, ten scans with a scan interval of 1 sec were accumulated. Each spectrum was recorded every 30 sec. The resolution of the spectra was 1 cm⁻¹.

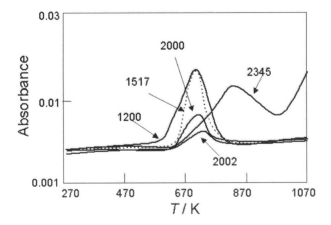

Figure 5-48. Absorbance of characteristic peaks of gas evolved by thermal decomposition of KL.

6.2   Kinetics of thermal decomposition of lignin

TG has extensively been used to analyze kinetic parameters of thermal decomposition of polymeric materials. In order to obtain activation energy, and reaction number, a differential method represented by the Freeman-Carroll method [5] has been utilized. However, the above method can be applicable in the cases, when the amount of reacted fraction x and rate of reaction dx/dt are directly measured. When we analyze the random scission of polymeric materials, the above conditions are not attained. In order to solve the above problem, the Ozawa-Flynn method has been introduced [6,7].

$$c = f(x) \tag{5.1}$$

When the decomposition is assumed to be explained in Arrehnius form, the following equation is obtained

$$\frac{dx}{dt} = A \exp\left(-\frac{\Delta E}{RT}\right) g(x) \tag{5.2}$$

where T is absolute temperature, R gas constant and E activation energy

$$\int_0^x \frac{dx}{g(x)} = A \int_0^\tau \exp\left(-\frac{\Delta E}{RT}\right) dt \tag{5.3}$$

$$\tau = \int_0^\tau \exp\left(-\frac{\Delta E}{RT}\right) dt \tag{5.4}$$

$$\tau = \frac{\Delta E}{\phi R}\left[p\left(\frac{\Delta E}{RT}\right)\right] \tag{5.5}$$

$$\log p(y) \equiv -2.315 - 0.4567 y$$

$$\log \phi_1 + 0.4567 \frac{\Delta E}{RT_1}$$

$$= \log \phi_2 + 0.4567 \frac{\Delta E}{RT_2} = \ldots\ldots$$

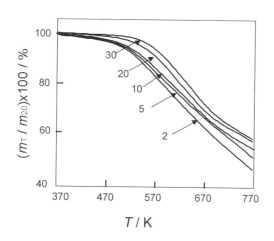

Figure 5-49. TG curves of dioxane lignin measured at various heating rates. Numerals in the figure show heating rate K min-1, Samples; DL was prepared from Incense ceder *(Libocedrus decurrens)*. Measurements; TG-DTA (Seiko Instruments), sample mass = ca. 5 mg, Pt pan, N2 flow rate = 100 ml min-1, temperature range; 300 –730 K.

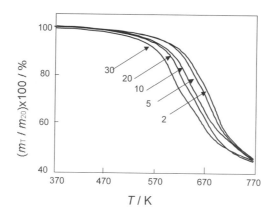

Figure 5-50. TG curves of acetylated dioxane lignin measured at various heating rates. Numerals in the figure show heating rate K min[-1] Samples; DL was acetylated by using unhydrous acetic acid in pirizine. Relative acethylation rate was 100 %. Measurement conditions, Figure 5-49 caption.

Figure 5.49 shows TG curves of dioxane lignin (DL) measured at various heating rates. As shown in Figure 5-49, two step decomposition is observed. In contrast, when the hydroxyl groups are converted into acetyl groups, the first step decomposition disappears as shown in Figure 5-50.

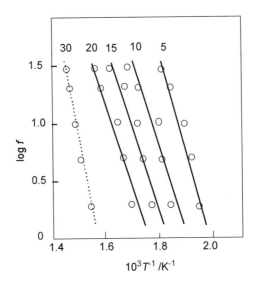

Figure 5-51. Relationship between log β and reciprocal temperature at various mass residue of DL. Numerals in the figure show amount of decomposition (%).

Assuming the decomposition reaction is governed by a single reaction process, the integral method was applied. Figure 5-51 shows the relationship between log β (β = dT/dt) and reciprocal temperature at the same mass residue of DL. Similar relationships were established for acetylated samples. When the mass residue is smaller than 20 %, lines shown in Figure 5-51 are parallel and this suggests that the integral method is applicable for the first step decomposition. From the gradient, activation energy was calculated as 136 kJ mol^{-1} for DL and acetylated DL (200 %) was 161 kJ mol^{-1}, respectively.

REFERENCES

1.   Sarkanen K. V. and Ludwig C. H., 1971, *Lignins, Occurance, Formation, Structure and Reactions,* Wiley-Interscience, New York.
2.   Lin S. Y.; Dence C W., *Methods in Lignin Chemistry,* Springer-verlag Berlin, 1992.

3. Hu, T. Q., *Chemical Modification, Properteis and Usage of Lignin*, Kluwer Acadimic/Plenum Pb., New York, 2002.
4. Nakano, J., ed. *Chemistry of Lignins*, Uni-Pub., Tokyo (Japanese).
5. Glasser W. G. and Northey R. A., Schultz T. P., *Lignin:Historical, biological and materials perspectives*, ACS symp. Ser. 742, 2000m ACS Washington, DC.
6. Hatakeyama T. and Hatakeyama; H., 1982, Temperature dependence of x-Ray diffractograms of amorphous lignins and polystyrenes", Polymer, 23, 475.
7. Hatakeyama, T., 1982, X-ray analysis of liquid to liquid transition of polystyrene. J. Macromol. Sci.,-Phys. B21(2), 299-305.
8. Hatakeyama, T., Nakamura K. and Hatakeyama; H., 1978, Differential thermal analysis of styrene derivatives related to lignin", *Polymer*, **19**, 593.
9. Ferry, J. D., 1961, *Viscoelastic Properties of Polymers*, John Wiley, New York.
10. Nielsen, L. W., 1962, *Mechanical Properties of Polymers*, Reinhold Pub., London.
11. Dawkins, J. V., ed., 1978, *Developments in Polymer Characterization-1*, 1980, Part -2. Applied Science Pub., London
12. Wada. Y., 1981, *Physical Properties of Polymers in Solid State*, Baifukan Pub. Tokyo (Japanese).
13. Van Krevelen, D.W.,1990, *Properties of Polymers*, 3rd ed., Elesevier, Amsterdam.
14. Matsuoka, S., 1992, *Relaxation Phenomena in Polymers*, Hanser Verlag, Munich, 1995 Japanese transl. by Ishihara, S., Kodansha Pub. Tokyo.
15. Hatakeyama T. and Quinn F. X., 1994, *Thermal Analysis.*: John Wiely, Chichester.
16. Turi, E. A. Ed. 1997 *Thermal Characterization of Polymeric Materials*. 2nd ed Academic Press, Orlando.
17. Hatakeyama T. and Liu, Z., 1998, *Handbook of Thermal Analysis*, John Weily, Chichester.
18. Liu, Z., Hatakeyama T. and Zhang X., 2001, *Thermal Measurements of Polymeric Materials* Industrial Chemistry Press, Beijing, (Chinese)
19. Hatakeyama, T., Nakamura, K. and Hatakeyama; H., 1978, Differential Thermal Analysis of Styrene Derivatives Related to Lignin, *Polymer*, **19**, 593-594.
20. Hatakeyama, H., Nakamura, K. and Hatakeyama, T., 1980, Studies on factors affecting the molecular motion of lignin and lignin-related polystyrene derivatives, *Tras. Pulp Paper Canada*, **81**, TR-105-TR110.
21. Hatakeyama, H. Nakano, J. Hatano A. and Migita N., 1969, Variation of infrared spectra with temperature for lignin and lignin model compounds, *Tappi*, **52**, 1724-1728.
22. Hatakeyama, T. and Serizawa, M., 1982, Glass transition of oligostyrene with different end groups, *Polym. J.,* **14**, 51-57.
23. Morck, R., Yoshida, H., Kringstad K. P. and Hatakeyama; H., 1986, Fractionation of kraft lignin by successive extraction with organic solvents: functional groups, ^{13}C-NMR Spectra and Molecular Weight Distributions, *Holzforschung*, **40**, Suppl. 51.
24. Yoshida, H., Morck, R., Kringstad K. P. and Hatakeyama; H,. 1987, Fractionation of kraft lignin by successive extraction with organic solvents,. Thermal properties of kraft lignin fractions", *Holzforschung*, **41**, 171.
25. Hatakeyama, H., Iwashita, K., Meshitsuka, G. and Nakano, J. 1975. Effect of Molecular Weight on Glass Transition Temperature of Lignin", *Mokuzai Gakkaishi*, **21**, 618.
26. Hatakeyama H. and Hatakeyama, T., 1972, Thermal Analysis of Lignin by Differential Scanning Calorimetry, *Cellulose Chem.*, **6**, 521-529.
27. Hatakeyama, T., Hirose S. and Hatakeyama, H., 1983, Differential scanning calorimetric studies on bound water in 1.4-dioxane aidolysis lignin; *Makromol. Chem.* **184**, 1265 - 1274.

28. Hatakeyama, T., Kanetusna, H. and Ichihara, S, 1989, Thermal analyis of polymer samples by round robin method. Part III. Heat capacity measurement by DSC, *Thermochimica Acta.*, **146**, 322-316.

29. Hatakeyama, T. Nakamura K., Hatakeyama; H. 1982, Studies on heat capacity of cellulose and lignin by differential scanning calorimetry, *Polymer,* **23**, 1801-1804.

30. Struik, L. C. E., 1978, *Physical Aging in Amorphous Polymers and Other Materials*, Elsevier, Amsterdam.

31. Chartoff, R P., 1997, Thermoplastic polymers, in Turi, E. A. ed. *Thermal Characterization of Polymeric Materials*. 2nd ed. Chapter 3, Academic Press, San Diego.

32. Hatakeyama, T. Nakamura, K. and Hatakeyama; H., 1978, Differential Thermal Analysis of Styrene Derivatives Related to Lignin, *Polymer,* **19**, 593-594.

33. Nakamura, K., Hatakeyama, T. and Hatakeyama; H., 1981, Differential scanning calorimetric studies on the glass transition temperature of polyhydroxystyrene Dderivatives containing sorbed water", *Polymer*, **22**, 473-476.

34. Nakamura, K., Hatakeyama, T. and Hatakeyama; H., 1981, Effect of water on the glass transition of styrene-hydroxystyrene copolymers, *Kobunshi Ronbunsyu*, **38**, 763-767 (Japanese).

35. Nakamura, K., Hatakeyama, T. and Hatakeyama; H., 1982 Relationship between hydrogen bonding and sorbed water in styrene- hydroxystyrene copolymers, *Kobunshi Ronbunsyu*, **39**53-58, (Japanese).

36. Nakamura, K., Hatakeyama, T. and Hatakeyama; H., 1982, The relationship between mechanical properties and hydrogen bonding of polyhydroxystyrene derivatives, *Kobunshi Ronbunsyu*, **39** (12), 765-770, (Japanese).

37. Nakamura, K., Hatakeyama, T. and Hatakeyama; H., 1983 Effect of substituent groups on hydrogen bonding of polyhydroxystyrene Derivatives", *Polym. J.*, **15**, 361 366.

38. Nakamura, K., Hatakeyama, T. and Hatakeyama; H., 1983, Relationship between hydrogen bonding and bound water in polyhydroxystyrene derivatives, *Polym.* **24**, 682-688.

39. Nakamura, K., Hatakeyama, T. and Hatakeyama; H., 1986, DSC studies on hydrogen bonding of poly(4-hydroxy-3, 5-dimethoxystyrene) and related derivatives, *Polym, J.*, **18**, 219-225.

40. Hatakeyama, H., Hayashi, E. and Haraguchi, T., 1977, Biodegradation of poly(3-methoxy-4-hydroxy stryrene), *Polymer*, **18**, 759

41. Kimura, M. Hatakeyama H. and Nakano, J., 1975, Torsional braid analysis of soft wood lignin and its model polymers", *Mokuzai Gakkaishi*, 21, 624.

42. Yano, S., Nakamura, K., Hatakeyama, T.. Hatakeyama; H., 1984, Temperature dependence of the tensile properties of lignin/paper composites, *Polym.* **25**, 890-893.

43. H. Hatakeyama and J. Nakano, 1970, Nuclear magnetic resonance studies on lignin in solid state, *Tappi*, 53, 472-475.

44. Hatakeyama H., and Nakano; J. 1970, Electron spin resonance studies on lignin and lignin model compounds, *Cellulose Chem. Technol.*, 4, 281.

45. Yano, S., Rgidahl, M., Kolseth, P. and deRuvo, A, 1984, Water-induced softening of lignosulfonates, Part I Effect of molecular mass, *Swnsk Papperstidning*, **87** (18) 170-176.

46. Yano, S., Rgidahl, M., Kolseth, P. and deRuvo, A, 1985, Water-induced softening of lignosulfonates, Part II Effect of the counterions, *Swnsk Papperstidning*, **88**, 10-14.

47. Hatakeyama, H., Hirose, S., Hatakeyama; T., 1989, Differential scanning calorimetry and NMR studies on the water-sodium lignosulfonate system, in *Lignin, Properties and Materials,* Glasser W., Sarkanen, eds ACS Symposium Series **397**, 274 –283.

48. Nguyen, T., Zavarin, E., Barrall II, E. W., 1981, Thermal analysis of lignocellulosic materials , Part I, unmodified , materials, *J.Macromol. Sci., Rev Macromol. Chem..*, **C20** 1-65.

49. Nguyen, T., Zavarin, E. and Barrall II, E. W., 1981,Thermal analysis of lignocellulosic materials, Part II, modified materials, J. Macromol. Sci., Rev Macromol. Chem., **C21** 1-60.

50. Jakab, E., Faiz, O., Till, F., and Sz'ekely, T., 1995, Thermogravimetry/mass spectrometry study of six lignins within the scope of an international round robin test, *J. Analytical. Applied Pyrolysis*, **35**, 167-179.

51. Caballero, J. A.. Font, R. and Marcilla, A., 1997, Pyrolysis of kraft lignin: yields and correlations, *J. Analytical. Applied Pyrolysis,* **39**, 161-183.

52. Jakab, E.. Faiz, O. and Till, F., 1997, Thermal decomposition of milled wood lignins studied by thermogravimetry/mass spectrometry, *J. Analytical. Applied Pyrolysis*, **40-41**, 171-186.

53. Hirose S.. Hatakeyama H, 1986,A Kinetic Study on Lignin Pyrolysis Using the Integral Method, *Mokuzai Gakkaishi*, **32**, 621

Chapter 6

PCL DERIVATIVES FROM SACCHARIDES, CELLULOSE AND LIGNIN

1. POLYCAPROLACTONE DERIVATIVES FROM SACCHARIDES AND CELLULOSE

Saccharide- and cellulose-based polycaprolactone (PCL) derivatives are obtainable by grafting PCL chains to the OH group of saccharides, cellulose and cellulose acetates having a degree of substitution (DS) below 3. Saccharide-based PCL's [1] have been synthesized by the polymerization of ε-caprolactone (CL) which was initiated by each OH group of glucose, fructose and sucrose. The amount of CL was varied from 1 to 5 moles per OH group of each saccharide. The polymerization was carried out for 12 hr at 423 K with the presence of a small amount of catalyst, dibutyltin dilaurate (DBTDL).

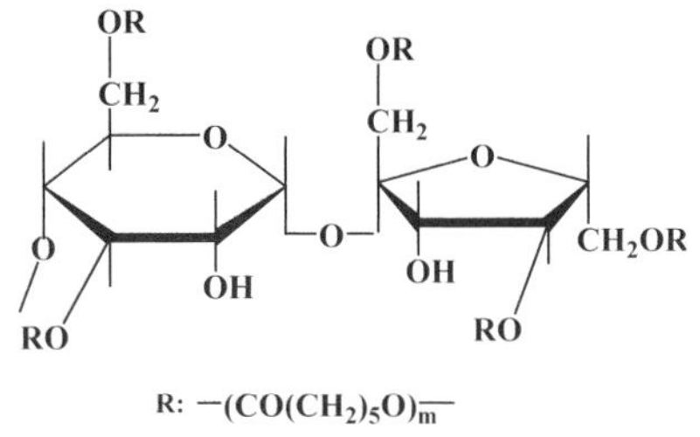

Figure 6-1. Schematic chemical structure of sucrose-based PCL's.

Figure 6.1 shows the schematic chemical structure of sucrose-based PCL's. The results of the characterization of glucose- (Glu-), fructose- (Fru-) and sucrose- (Suc-) based PCL's have been reported elsewhere [2].

1.1   Polycaprolactone derivatives from cellulose acetate (CA)

Cellulose acetate (CA) - based polycaprolactone derivatives (CAPCL's) were synthesized from CA which was commercially obtained from Kodak Co. Ltd. The specifications of CA were as follows: acetyl content, 39.87 %; $M_w = 6.32 \times 10^4$; $M_w/M_n = 2.27$. Distilled CL, which was dehydrated in benzene by reflux method, was added to dried CA and polymerization was carried out for 22 hrs at 423 K with the presence of a small amount of DBTDL. CL/OH (mol mol^{-1}) ratios were varied from 2 to 20 (mol mol^{-1}): (CL/OH ratio = 2, 5, 8, 10, 15, 20). When the samples with CL/OH ratio of 2 and 5 were synthesized, N-methyl-2-pyrorydone was used as a solvent in order to make the reaction go smoothly. When CL/OH ratio was over 8, it was unnecessary to use pyrorydone, since CL itself worked as a solvent of CA. The obtained CA-PCL's were dissolved in hot acetone and then put in methanol dropwise in order to precipitate purified CA-PCL's. Precipitates in flake shape were obtained. Samples were dried in an oven in vacuum at 378 K for 12 hrs. Figure 6-2 shows a schematic chemical reaction for the synthesis of CAPCL.

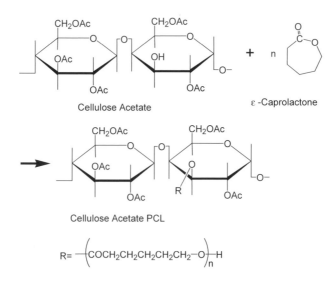

Figure 6-2. Schematic chemical reaction for the synthesis of CAPCL [2].

1.2   Polycaprolactone derivatives from cellulose

Cellulose-based polycaprolactone derivatives (CellPCL's) have been synthesized by 2 step reactions [3]. In the first step, cellulose powder was suspended in N, N-dimethylacetoamide (DMAc) and cellulose soaked with DMAc was obtained by filtration. The above cellulose soaked with DMAc was dissolved in DMAc solution of lithium chloride (LiCL), dehydrated by refluxing with benzene and then reacted with CL with the presence of a small amount of catalyst, triethylamine. CL/OH (mol mol^{-1}) ratio was 0.33 at this first stage. The above obtained CellPCL was dissolved in 2-methyl-N-pyrolydone. Distilled CL was added to the above solution and polymerization was carried out for 22 hrs with the presence of a small amount of DBTDL. CL/OH (mol mol^{-1}) ratios were varied from 0.66 to 5 (mol mol^{-1}). Precipitates in flake shape were obtained by putting the above obtained DMAc solution of CellPCL's into methanol. Figure 6-3 shows the schematic chemical structure of cellulose-based PCL.

$$R= \left(\overset{\overset{\displaystyle O}{\|}}{C}CH_2CH_2CH_2CH_2CH_2\right)_n O\text{-}H$$

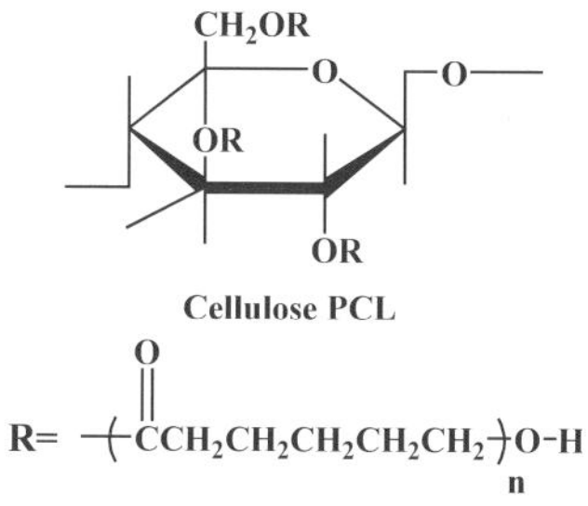

Figure 6-3. Schematic chemical structure of cellulose-based PCL [2].

1.3   Thermal properties

1.3.1   Cellulose acetate-based PCL derivatives

Figure 6-4 shows stacked DSC heating curves of CA and CAPCL's measured at 10 K min^{-1}. As shown in Figure 6-4, a baseline deviation showing glass transition is clearly observed for the CA sample (shown as CL/OH ratio = 0) and CAPCL samples. DSC curves were also measured at various heating rates from 2 to 40 K min^{-1} and the heating rate dependency of T_g was clearly observed. It is known that glass transition temperature (T_g) of CA depends mainly on acetyl contents and not on molecular weight or molecular weight distribution [4]. T_g of CA observed in this study was 420 K and this value accords well with reported values [5]. Both glass transition and melting can be observed in the DSC curve of PCL. A melting peak of

CAPCL was observed at around 323 K. A melting peak of PCL was observed at 340 K. This value accords well with the reported values [6]. Accordingly it may be reasonable to consider that the melting peaks observed in Figure 6-4 correspond to the melting of the PCL side chain grafted to the OH group of CA [2].

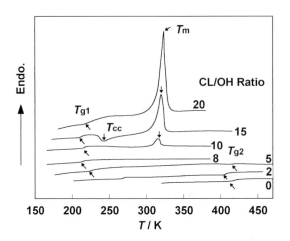

Figure 6-4. Stacked DSC heating curves of CA and CAPCL's. Numerals shown in the figure indicate CL/OH ratios. T_g: glass transition temperature, T_{cc}: cold crystallization temperature, T_m: melting temperature, Measurements; heat-flux type DSC (Seiko Instruments), sample mass = ca. 5mg, heating rate = 10 K min^{-1}, N$_2$ flow rate = 30 ml min^{-1}.

Figure 6-5 shows change of T_g with CL/OH ratio in CAPCL [2]. It is known that dry cellulose shows no glass transition in a temperature range from 293 K to thermal decomposition temperature [6]. The results of heat capacity (C_p) measurement of dry cellulose showed that the gradient of C_p increase depends on the crystallinity of cellulose [6]. The T_g of CA is observed as shown in Figure 6-5. The above facts suggest that T_g of cellulose can be changed and observed by the introduction of large side chain molecules, such as acetyl group and caprolactone chain. Figure 6-5 shows that T_g of CA (T_{g2}) is observed in the initial stage and becomes difficult to detect when CL/OH ratio exceeds 15 mol mol^{-1}. It is considered that T_g of cellulose chain is observable when intermolecular distance expands by the introduction of large side chain molecules, and the geometrical free space enhances the main chain motion. T_g of PCL (T_{g1}) part of CAPCL decreases in the initial stage and increases slightly after reaching a minimum point at around CL/OH = 10 mol mol^{-1}. This T_g increase observed in the sample with CL/OH ratio over 10 suggests that the molecular motion of PCL random chains is restricted by the presence of crystalline region.

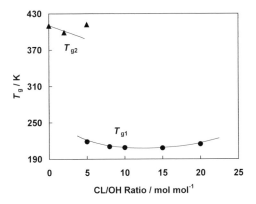

Figure 6-5. Change of T_g with CL/OH ratio in CAPCL.

As shown in Figure 6-4, the melting peak of PCL chains is observed for the PCL samples with CL/OH ratio 10, 15 and 20 mol mol^{-1}. A broad exothermic peak due to cold crystallization (T_{cc}) at around 240 K is observed for the sample with CL/OH ratio of 15 mol mol^{-1}.

In order to obtain samples having the same thermal history in DSC measurements, the samples were heated to 393 K and quenched to 123 K. The heating run was carried out at 2 K min^{-1}. With heating slowly, T_m shifted to the high temperature side and ΔH_m (J g^{-1}) increased. T_m of the sample with CL/OH ratio 8 mol mol^{-1} became observable. However, T_{cc} was hardly observed in the samples heated at 2 K min^{-1}.

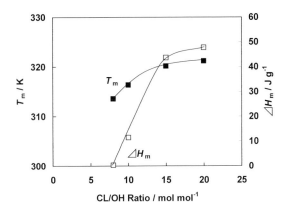

Figure 6-6. Change of T_m, ΔH_m with CL/OH ratio in CAPCL.

Figure 6-6 shows the change of T_m, ΔH_m with CL/OH ratio. Temperature of melting (T_m) and enthalpy of melting (ΔH_m) increase with increasing CL/OH ratio. Both T_m and ΔH_m of the samples heated at 2 K min^{-1} are higher than those heated at 10 K min^{-1}. The obtained results indicate that the higher order structure formation is strongly affected by thermal history, i.e. the crystallinity of PCL chains increases during heating.

The heat capacity difference at T_g (ΔC_p) varies inversely with T_g and the highest value of ΔC_p is observed at around CL/OH ratio 10 mol mol^{-1}, as shown in Figure 6-7. The variation of T_g suggests that the main chain motion of PCL is enhanced with increasing chain length. At the same time, it is suggested that the molecular motion of PCL chain of CAPCL is restricted when long PCL chain molecules form a regular crystalline structure.

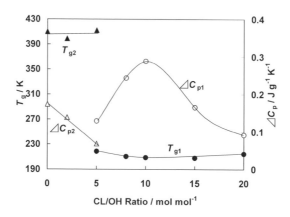

Figure 6-7. Change of T_g, ΔC_p with CL/OH ratio in CAPCL.

A reciprocal relationship is established between T_g and ΔC_p values among a large number of amorphous polymers when the samples are completely random and have no intermolecular bonding [7]. When the relationship between T_g and ΔC_p is considered for complex polymers, such as CAPCL, it must be taken into consideration that the thermodynamically equilibrium state is not attained in either glassy or rubbery state when ΔC_p values are measured by dynamic measurement.

The crystalline region is formed in the samples with CL/OH ratio over 8, as shown in Figures 6-6 and 6-7. A part of the amorphous region is transformed to the crystalline region that can be observed as T_{cc}. This suggests that molecular motion at a temperature higher than T_g of the samples with high CL/OH ratio is restricted and accordingly ΔC_p becomes small. It is also known that ΔC_p values cannot be estimated for the samples heated at 2 K min^{-1}. This indicates that the number of chain molecules contributing to the enhanced molecular motion is reduced by slow heating. It should also be noted that a part of the main chain still forms a random structure, since a small baseline inflection can be observed when the sample is measured at the heating rate of 10 K min^{-1}.

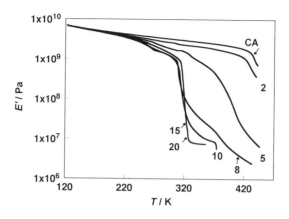

Figure 6-8. Representative E' curves of CAPCL with various CL/OH ratios (mol mol^{-1}). Numerals in the figure indicate CL/OH ratio [8]. Measurements; DMA (Seiko Instruments), heating rate = 2 K min^{-1}, sample = sheet, N$_2$ gas frequency =10 Hz [8].

Figure 6-8 shows representative E' curves of CAPCL with various CL/OH ratios measured at frequencies from 5 Hz. E' (Pa) gradually decreases from 120 to 220 K and a steep decrease is observed at temperatures over 320 K, depending on PCL chain length (CL/OH ratio).

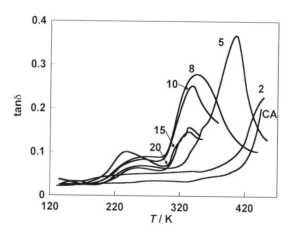

Figure 6-9. Representative tan δ curves of CAPCL with various CL/OH ratios (mol mol^{-1}). Numerals in the figure indicate CL/OH ratio [8].

Tan δ curves of CAPCL's with CL/OH ratios to low temperature side, tan δ peaks were designated as α-dispersion, β-dispersion and γ-dispersion, respectively. The variation of E'' and tan δ corresponds to the number of CL/OH ratio. The α-dispersion becomes marked when CL/OH ratio is over 5. The β-dispersion corresponds to the molecular motion of amorphous chains of CAPCL. When the crystalline region is formed, it restricts the incoherent movement of random chains and the number of molecular chains involving enhanced molecular motion decreases. Tan δ values at the peak temperature with CL/OH ratio show a maximum at around CL/OH = 5 to 10 mol mol^{-1} [8]. The variation of the intensities of tan δ at the β- dispersion is similar to that of ΔC_p shown in Figure 6-7. Both results indicate that the molecular mobility of the main chains is high at around CL/OH ratio = 5 to10 mol mol^{-1}. When the side PCL chain length is short or the crystallinity increases, free molecular motion is restricted.

Activation energy (E_a) of each dispersion was calculated from frequency dependency of maximum temperature of tan δ assuming the applicability of Arrehenius equation [8]. E_a of α-dispersion is shown in Figure 6-10. E_a reaches a minimum at around CL/OH ratio = 8 to 10 mol mol^{-1}. As already reported, E_a of the main chain motion ranges from 100 J mol^{-1} to 250 J mol^{-1} [7]. Accordingly, it is appropriate to consider that α-dispersion is attributed to the main chain motion of the PCL chain.

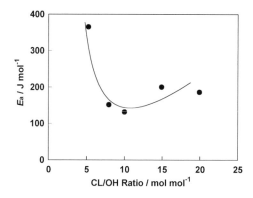

Figure 6-10. Change of E_a (activation energy of α-dispersion) of CAPCL with CL/OH ratio [8].

As described above, it can be said that the main chain motion of both CA and PCL chains is observed in a CL/OH ratio 2 ~10 mol mol^{-1}. Melting of the crystalline region of PCL chains can be observed at around 320 K. The main chain motion of CA is observable when CL/OH ratio is low (0 ~ 5 mol mol^{-1}). However, it becomes difficult to detect the main chain motion of CA when CL/OH ratio exceeds 8. Variation of T_g, temperature of β-dispersion of PCL and ΔC_p indicates that the molecular chain of CAPCL becomes mobile with increasing CL/OH ratio until CL/OH reaches 10 mol mol^{-1}. When

CL/OH ratio is over 15, the main chain motion of PCL is restricted by crystal growth. The sample with CL/OH ratio 5 ~ 10 mol mol^{-1} shows intermediate characteristics. Amorphous structure is formed by quenching CAPCL from the molten state. Cold crystallization is observed during heating CAPCL with CL/OH ratio = 5 ~ 10 mol mol^{-1}. It is also suggested that the crystalline region is formed, when the sample is slowly heated. This indicates that the degree of freedom of molecular chains decreases with crystal growth. By the comparison of DSC and DMA results, it can be said that molecular motion measured by DSC corresponds to that measured at low frequency range by DMA.

TG and DTG curves of CAPCL's with various CL/OH ratios are shown in Figure 6-11. The samples with a low CL/OH ratio from 0 to 5 mol mol^{-1} decompose in one stage and peak temperature of DTG is observed at around 650 K. When the CL/OH ratio exceeds 8 mol mol^{-1}, two peaks are observed at around 620 and 700 K. Mass residue (MR), [(m_T /m_{293}) x 100] at around 720 K, where m_t is mass at T K and m_{20} mass at 293 K, decreased markedly with increasing CL/OH ratio, although it is not shown in the figure.

As seen from DTG curves shown in Figure 6-11, the DT_{d1} peak temperature maintains an almost constant value with increasing CL/OH ratio.

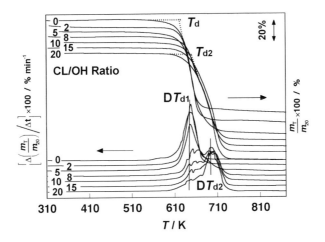

Figure 6-11. TG and DTG curves of CAPCL's with various CL/OH ratios [2]. m_{20}: mass at 293K, m_T: mass at T. Measurements; TG-DTA (Seiko Instruments TG/DTA 220), heating rate = 20 K min^{-1}, sample mass = ca. 5 mg, N$_2$ flow rate = 10 K min^{-1}.

The DT_{d2} peak temperature is also almost constant as shown in Figure 6-11. At the same time, the height of DT_{d1} peak temperature markedly decreases at CL/OH ratio below 8.

Figure 6-12 shows the relationship between T_d and CL/OH ratio in CAPCL's. This figure clearly shows that the thermal degradation of CAPCL's proceeds in 2 steps. Thermal degradation at lower temperature is observed at around 620 K and that observed at higher temperature is observed at around 660 K. This result suggests that the 2 step thermal degradation of CAPCL's is probably caused by two chemically different structures: CA and PCL structures in CAPCL. Change of DT_d with CL/OH ratio in CAPCL shown in Figure 6-13 supports the above results since two markedly different DT_d's are observed at around 640 and 700K.

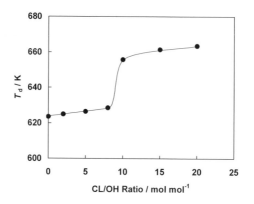

Figure 6-12. Change of T_d with CL/OH ratio in CAPCL [2].

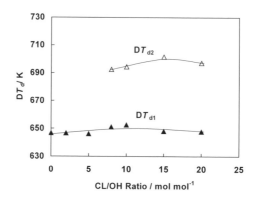

Figure 6-13. Change of DT_d with CL/OH ratio in CAPCL.

Figure 6-14 shows the relationship between *MR* and CL/OH ratio at 730 K. As seen from the figure, *MR* decreases markedly with increasing CL/OH ratio. This suggests that PCL chains degrade easily compared with CA chains, although T_d of PCL chains considered to be higher than that of CA chains.

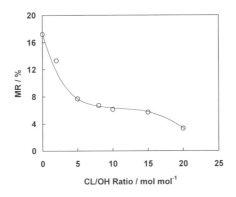

Figure 6-14. Change of *MR* with CL/OH ratio in CAPCL at 730 K.

Figure 6-15 shows a representative stacked three-dimensional diagram showing the change of IR intensity, wavenumber and temperature in TG-FTIR of CAPCL with CL/OH ratio = 8 mol mol^{-1}. Wavenumber ranges from 600 to 4000 cm^{-1} and temperature ranges from 310 to 980 K. As seen from the diagram, IR absorption bands can mainly be observed in the temperature range from 523 to 723 K, where 1160 cm^{-1} corresponds to vC-O-, 1260 cm^{-1} to -C (=O)-O-C-, 1517 and 1617 cm^{-1} to vC=C, 1770 cm^{-1} to vC=O, 2345 cm^{-1} to vCO$_2$, 2945 cm^{-1} to vC-H and 3700 cm^{-1} to vOH.

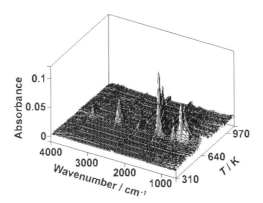

Figure 6-15. Stacked three-dimensional TG-FTIR diagram for CAPCL (CL/OH ratio = 8 mol mol. Measurements; TG-DTA-FTIR (Seiko Instruments TG /DTA220, SSC5200, FTIR, Japan Spectroscopic Co. Ltd., FT/IR 700), heating rate = 20 K min^{-1}, sample mass = ca. 7 mg, gas flow rate = 100 ml min^{-1}, temperature of gas transfer system = 540 K, resolving of FTIR = 8 cm^{-1}, number of integration = 10, data incorporation time = 30 s.

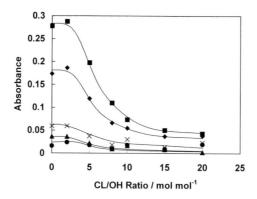

Figure 6-16. Change of IR peak intensity corresponding to C-O-C, C=O, CO$_2$, CH, OH bands with CL/OH ratio in PCL derivatives of cellulose acetate at 653 K [2]. ◆C-O-C, ■C=O▲ CO$_2$, ●CH, ✕OH.

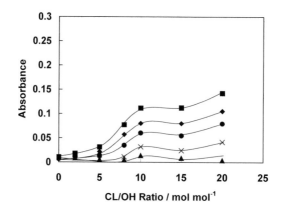

Figure 6-17. Change of IR peak intensity corresponding to C-O-C, C=O, CO$_2$, CH, OH bands with CL/OH ratio in CAPCL at 703 K [2].

Figures 6-16 and 6-17 show the changes of C-O-C, C=O, CO$_2$, CH, and OH peak intensities with CH/OH ratios in CAPCL's at 653 and 703 K. The above temperatures correspond to the temperatures where DT_{d1} and DT_{d2} are observed. As seen from Figure 6-16, the changes of the peak intensities corresponding to C-O-C, C=O and OH absorption peak at 653 K decrease with increasing CL/OH ratios. This suggests that the evolved gases are mainly formed by the degradation of the cellulose acetate part of CAPCL's.

On the other hand, as seen from Figure 6-17, the peak intensities of the C-O-C, C=O, CH and OH bands increase with increasing CL/OH ratios at 703 K. This indicates that the evolved gases are formed at 703 K by the thermal degradation of PCL chains, because the C-O-C, C=O and CH absorption peaks correspond to the structure of PCL chain, as can be inferred from the marked increase of the peak intensities with increasing CL/OH ratios.

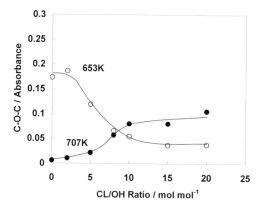

Figure 6-18. Change of IR peak intensity corresponding to C-O-C band with CL/OH ratio in CAPCL at 653 and 703 K.

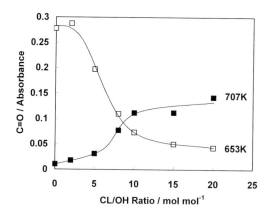

Figure 6-19. Change of IR peak intensity corresponding to C=O band with CL/OH ratio in CAPCL at 653 and 703 K.

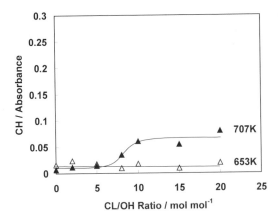

Figure 6-20. Change of IR peak intensity corresponding to CH band with CL/OH ratio in CAPCL at 653 and 707 K.

The relationships shown in Figures 6-16 and 6-17 are more easily understandable with reference to the relationship shown in Figures 6-18, 6-19 and 6-20. As seen from Figures 6-18 and 6-19, IR peak intensities corresponding to C-O-C and C=O bands decrease with increasing CL/OH ratio at 653 K, while both peak intensities markedly increase with CL/OH ratio at 703 K. Accordingly, it is considered that both bands strongly associate with PCL chains and that PCL chains do not degrade at 653 K, but degrade at 703 K. Peak intensity of CH band does not change with CL/OH ratio at 653 K, while it increases at 707 K although peak intensity is not strong. This result indicates that the degradation of the molecular structure corresponding to CH band probably does not occur at 653 K.

1.3.2   Cellulose-based PCL's

As shown in Figure 6-3, the chemical structure of CellPCL is different from that of CAPCL, since each OH group at the glucose unit of cellulose is almost converted to PCL chains. On the other hand, CAPCL shown in Figure 6-20 has about one PCL chain per cellobiose unit of the cellulose chain.Figure 6-21shows stacked DSC heating curves of CellPCL's measured at the heating rate of 10 K min^{-1}. As seen from Figure 6-21, a baseline deviation showing glass transition is not observed for the cellulose sample (shown as CL/OH ratio = 0), while T_g's for CellPCL's are clearly observed. CellPCL's with CL/OH ratios = 0.33 and 0.66 show T_{g2} at around 420 K. T_{g2} is considered to be related with the main chain motion of the cellulose backbone of CellPCL, since this motion was also observed in the case of CAPCL, as described at section 1.3.1 in this chapter. T_{g1}'s are observable when CL/OH ratios are between 0.3 and 5. However, T_g is difficult to observe when CL/OH ratio exceeds 6, since endothermic peaks caused by melting of PCL (T_m) make it difficult to detect the baseline deviation caused by glass transition.

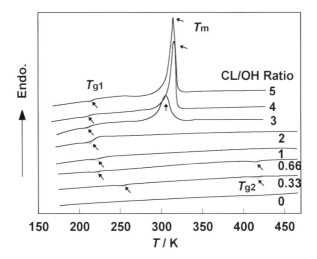

Figure 6-21. DSC heating curves of CellPCL [2]. Measurements; heat-flux type DSC (Seiko Instruments), heating rate = 0 K min^{-1}, N$_2$ gas flow rate = 100 ml min^{-1}, sample mass = ca. 7 mg.

Figure 6-22 shows the change of T_g and ΔC_p with CL/OH ratio. As seen from Figures 6-2 and 6-3, the numbers of PCL side chains attached to each cellobiose unit of the cellulose backbone are almost 6 times that of CAPCL. Accordingly, T_g's of CellPCL shown in Figure 6-22 are almost similar to those of CAPCL shown in Figure 6-5, although CL/OH ratio of CellPCL is far lower than that of CAPCL. The above results shown in Figure 6-22 are reasonable, since it is considered that T_g becomes observable as the baseline deviation of DSC heating curves, when intermolecular distance expands by the introduction of large side chain molecules such as PCL and when the geometrical free space enhances the main chain motion. T_{g1} of CellPCL decreases in the initial stage when CL/OH ratio is low and increases slightly after reaching a minimum point at around CL/OH = 3 mol mol^{-1}. This T_g increase observed in the case of CellPCL's with CL/OH ratio over 3 suggests that the molecular motion of PCL chains is restricted by the presence of the crystalline region. Change of ΔC_p in Figure 6-22 supports the above explanation, since ΔC_p increases with increasing CL/OH ratio when it is below 3 and after reaching a maximum ΔC_p decreases again with increasing CL/OH ratio. It is considered that the effect of the ordered arrangement of PCL chains on the molecular motion becomes marked and restricts the molecular motion of PCL chains in CellPCL, when the numbers of PCL chains are sufficiently large.

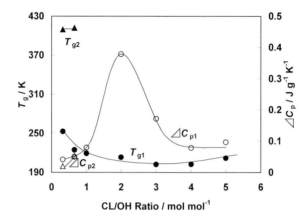

Figure 6-22. Change of T_g and ΔC_p with CL/OH ratio in CellPCL.

As shown in Figure 6-23, the melting of PCL chains is observed for the PCL samples with CL/OH ratio 3, 4 and 5 mol mol^{-1}. An endothermic peak observed at around 310 K is attributable to the melting of PCL chains.

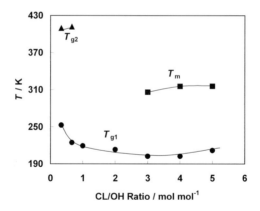

Figure 6-23. Change of T_g and T_m with CL/OH ratio in CellPCL.

Figure 6-24 also shows the change of T_m and ΔH_m with CL/OH ratio. T_m and ΔH_m increase with increasing CL/OH ratio. It is noteworthy that T_m and ΔH_m values almost level off when CL/OH ratio becomes 5. This result indicates that large numbers of PCL chains attached to the cellulose backbone have a strong influence on the molecular arrangement of CellPCL compared with that of CAPCL. It should be noted that the number of PCL side chains attached to each cellobiose unit in CellPCL is far larger than that attached to the acetylated cellobiose unit of CAPCL.

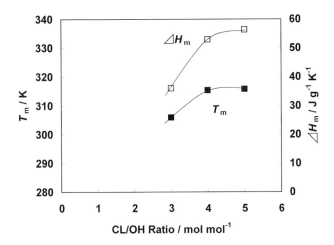

Figure 6-24. Relationship among T_m, ΔH_m and CL/OH ratio in CellPCL.

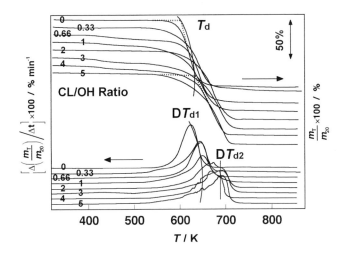

Figure 6-25. TG and DTG curves of CellPCL with various CL/OH ratios.

Figure 6-25 shows TG and DTG curves of CellPCL's with various CL/OH ratios from 0 to 5. The samples with a low CL/OH ratio below 1 mol mol^{-1} decompose in one stage and the peak temperature of DTG (DT_{d1}) increases from ca. 610 to 630 K. When the CL/OH ratio exceeds 2 mol mol^{-1}, the DTG peak (DT_{d2}) becomes observable in the temperature range between ca. 650 and 690 K. However, as seen from Figure 6-26, T_d does not show a clear 2 step thermal degradation as was observed in CAPCL.

Figure 6-27 shows the change of *MR* with CL/OH ratio in CellPCL at 730 K. As seen from the figure, *MR* shows a maximum when CL/OH ratio is 0.33 and then decreases with increasing CL/OH ratio. This may indicate that CellPCL is most stable against thermal degradation when cellulose structure is grafted by short PCL chains.

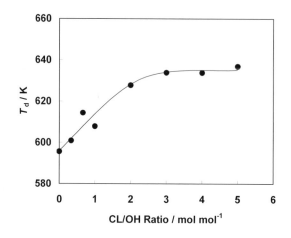

Figure 6-26. Change of T_d with CL/OH ratio in CellPCL.

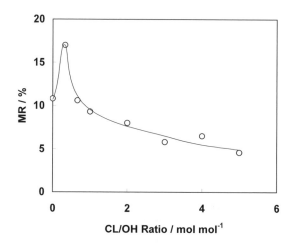

Figure 6-27. Change of *MR* with CL/OH ratio in CellPCL at 730 K.

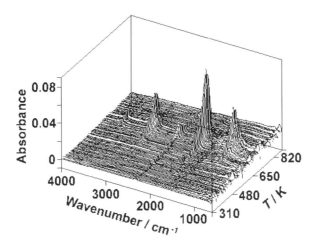

Figure 6-28. Stacked three-dimensional TG-FTIR curves of CellPCL (CL/OH ratio =2 mol/mol⁻¹).

Figure 6-28 shows a stacked three-dimensional diagram showing the change of IR intensity and wavenumber with temperature in TG-FTIR of CellPCL with CL/OH ratio 2 mol mol⁻¹. Wavenumber ranges from 600 to 4000 cm⁻¹ and temperature ranges from 310 to 870 K. As shown in the diagram, IR absorption bands are observed in the temperature range from 570 to 720 K. Most IR peaks observed for the samples are similar to those shown in Figure 6-15.

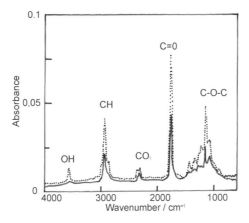

Figure 6-29. TG-FTIR of CellPCL (CL/OH ratio = 2 mol/mol) solid line: 643 K, broken line: 703 K.

FTIR spectra at 643 and 703 K for CellPCL with CL/OH ratio 2 mol mol^{-1} is shown in Figure 6-29. The assignments for the IR peaks are indicated in Figure 6-29. It is seen that the intensities of absorption peaks change according to the degradation temperature of the samples. At 643 K, absorption intensity of each IR peak is lower than that at 703 K. This suggests that the degradation of CellPCL at 643 K is less than that at 703 K.

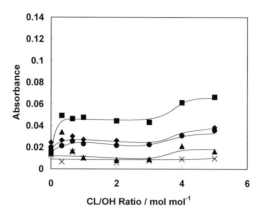

Figure 6-30. Change of IR peak intensity corresponding to C-O-C, C=O, CO$_2$, CH, OH bands with CL/OH ratio in CellPCL at 643 K. ◆: C-O-C, ■: C=O, ▲: CO$_2$, ●: CH, ×: OH.

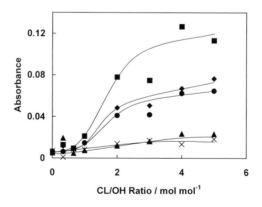

Figure 6-31. Change of IR peak intensity corresponding to C-O-C, C=O, CO$_2$, CH, OH bands with CL/OH ratio in Cell PCL at 703 K. ◆: C-O-C, ■: C=O, ▲:CO$_2$, ●:CH, ×:OH.

Changes of C-O-C, C=O, CO$_2$, CH, and OH peak intensities with CH/OH ratios in CellPCL's at 643 and 703 K are shown in Figures 6-30 and 6-31. Figure 6-30 shows the change of peak intensities corresponding to C-O-C, C=O, and CH absorption peaks at 643 K with increasing CL/OH ratios. As seen from Figure 6-31, the intensities of the C-O-C, C=O, CH, CO_2 and OH absorption peaks increase with increasing CL/OH ratios at 703 K. As clearly seen, the peak intensity corresponding to the C-O-C, C=O, and CH absorption increases markedly with increasing CL/OH ratios.

In comparison with the results obtained for CAPCL, it is appropriate to conclude that the degradation occurring at around 643 K corresponds to that of the cellulose structure in CellPCL, and that the degradation of CellPCL occurring at around 703 K corresponds to that of the PCL chain.

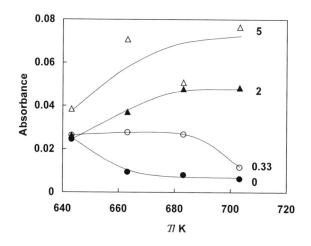

Figure 6-32. Change of IR peak intensity corresponding to C-O-C band (1160cm⁻¹) of CellPCL with temperature. Numerals show CL/OH ratio.

The results shown in Figure 6-32 accord well with those shown in Figure 6-27. Peak intensity of C-O-C band in CellPCL with Cl/OH ratio of 0.33 decreases in the temperature range over 700 K, while the IR peak intensity in CellPCL's with CL/OH of 2 and 5 increases with increasing temperature. The above results show that *MR* of CellPCL with CL/OH ratio of 0.33 reached a maximum at 730 K.

As seen from Figure 6-33, IR peak intensities corresponding to C-O-C band decrease with increasing CL/OH ratio at 643 K, while the peak intensity markedly increases with CL/OH ratio at 703 K. Accordingly, it is considered this C-O-C band associates with PCL chains and that PCL chains in CellPCL do not degrade at 643 K, although CellPCL degrades at 703 K. The above results strongly indicate the occurrence of the thermal degradation of PCL chains at around 700K.

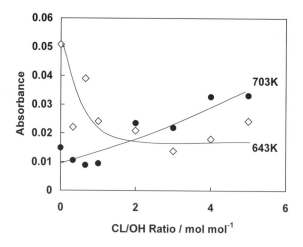

Figure 6-33. Change of IR peak intensity corresponding to C-O-C band with CL/OH ratio in CellPCL at 643 and 703 K.

2.   POLYCAPROLACTONE DERIVATIVES FROM LIGNINS

2.1   Lignin-based PCL's

Alcoholysis lignin-based PCL (ALPCL) [3] and kraft lignin-based PCL (KLPCL) [3] were synthesized by polymerization of CL which was initiated by the hydroxyl (OH) group of the above lignins. The amount of CL was

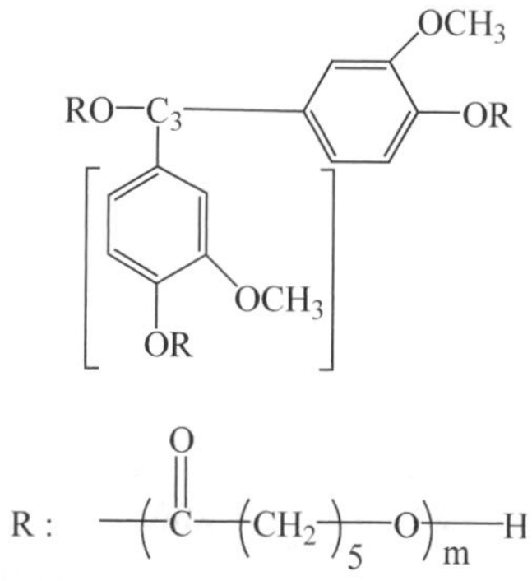

Figure 6-34. Schematic chemical structure of LigPCL

varied from 1 to 25 moles per OH group of each lignin (CL/OH ratio = 1, 2, 3, 4, 5, 10, 15, 20 and 25 mol mol^{-1}). The polymerization was carried out for 12 hr at 423 K with the presence of a small amount of DBTDL. ALPCL and KLPCL sheets were prepared by heat-pressing the synthesized polymers at 430 – 450 K under ca. 10 MPa. A schematic chemical structure of LigPCL is shown in Figure 6-34.

2.2 Thermal properties

Figures 6-35 and 6-36 show part of the stacked magnified DSC heating curves of ALPCL and KLPCL with various CL/OH ratios from 2 to 25 mol mol^{-1} measured in the temperature range from 173 to 423 K. A marked change in the baseline of DSC curves due to glass transition was observed. T_g's were determined by the method reported previously [9].

As shown in Figure 6-35, in the case of the DSC curve representing ALPCL with CL/OH ratio of 4 mol mol^{-1}, a prominent exothermic peak due to cold-crystallization of ALPCL and also a recognizable endothermic peak due to melting of crystals are observed. As seen from Figure 6-36, similar phenomena are observed for KLPCL with CL/OH ratio of 5 mol mol^{-1}. When CL/OH ratio is over 10 mol mol^{-1}, marked endothermic peaks due to melting are observed in ALPCL and KLPCL. The above results suggest that AL- and KLPCL's with CL/OH ratios over 10 mol mol^{-1} have a clear crystalline region in the molecular structure.

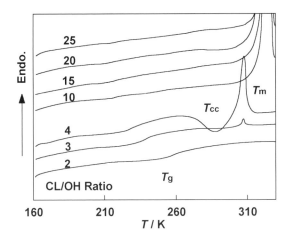

Figure 6-35. Stacked DSC heating curves of ALPCL with various CL/OH. Measurements; heat-flux type DSC (Seiko Instruments), heating rate = 10 K min^{-1}, sample mass = ca. 7 mg, N$_2$ gas flow rate = 100 ml min^{-1}.

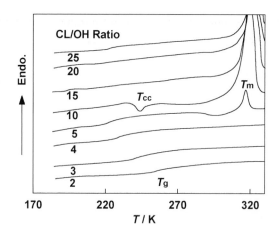

Figure 6-36. Stacked magnified DSC heating curves of KLPCL with various CL/OH [2].

Figure 6-37 shows the change of T_g's of AL- and KLPCL's. The T_g decreases markedly with increasing CL/OH ratio in the region where CL/OH ratio is below 10 mol mol^{-1} and then the T_g increases with increasing CL/OH ratio in the region where CL/OH ratio exceeds 10 mol mol^{-1}. The increase of T_g over CL/OH ratio = 10 mol mol^{-1} indicates that the crystalline region increases by the introduction of long PCL chains, and that this crystalline region restricts the motion of PCL chains in AL- and KLPCL's.

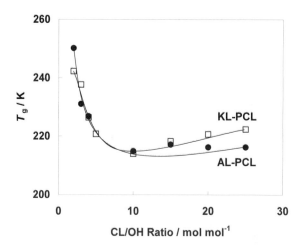

Figure 6-37. Change of T_g's with CL/OH ratio.

As strongly suggested by the above facts, the effect of the structural difference between AL and KL does not affect markedly the phase transition of AL- and KLPCL's, suggesting that both lignins are available as good raw materials for the preparation of PCL derivatives.

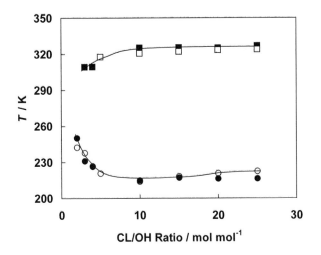

Figure 6-38. Change of T_g and T_m of KLPCL and ALPCL with CL/OH ratio. KLPCL, ●: T_g, ■: T_m; ALPCL, ○: T_g, □: T_m.

Figure 6-38 shows a phase diagram of KLPCL according to the above results. In general, no large difference is observed between T_g and T_m of KLPCL and those of ALPCL.

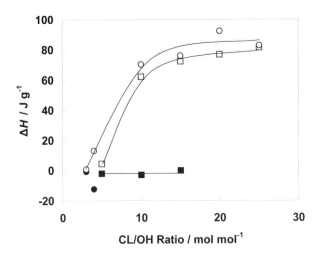

Figure 6-39. Change of ΔH's of T_{cc} and T_m with CL/OH ratio in AL- and KLPCL. KLPCL, ■: T_{cc}, □: T_m ; ALPCL, ●: T_{cc} , ○: T_m.

Figure 6-39 shows change of ΔH's of cold crystallization and melting with CL/OH ratio in AL- and KLPCL. Enthalpy of cold crystallization did not change with CL/OH ratio. Enthalpy of T_m, on the contrary, increases clearly with increasing CL/OH ratio and almost levels off when CL/OH ratio exceeds 10.

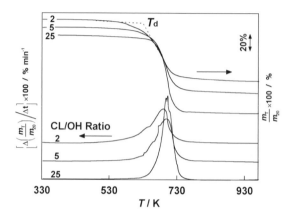

Figure 6-40. TG and DTG curves of KLPCL's with CL/OH ratios of 2, 5 and 25 [2].

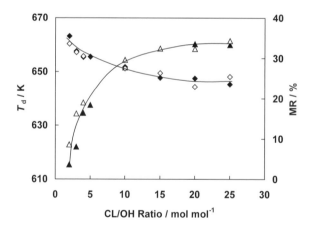

Figure 6-41. Change of T_d and MR with CL/OH ratio. KLPCL, ▲: T_d, ◆: MR ; A-PCL, △: T_d, ◇: MR.

TG and DTG curves of KLPCL's with CL/OH ratios of 2, 5 and 25 mol mol^{-1} are shown in Figure 6-40. Similar TG and DTG curves were also obtained for ALPCL's. DTG curves show that samples start to decompose at around 570 K. DTG curves of KLPCL's also show that samples with

CL/OH ratios of 2 and 5 mol mol^{-1} may decompose in multiple steps in the temperature range from 570 to 720 K, since DTG peaks are broad and show tailing to the low temperature side. The DTG curve for KLPCL with CL/OH ratio 25 mol mol^{-1} shows an obviously narrow curve without tailing. The above facts seem to suggest that the thermal degradation of PCL with lower values of CL/OH ratios may be affected by thermal degradation of the lignin core structure in KLPCL derivatives.

Figure 6-41 shows the change of T_d and *MR* with CL/OH ratio for AL- and KLPCL's. As seen from the figure, T_d initially increases and then levels off with increasing CL/OH ratio over 10 mol mol^{-1}. It has been reported that lignin becomes thermally stable when OH groups of lignin are methylated and acetylated [10], since reactive OH groups are blocked. Accordingly, the increase in T_d values of AL- and KLPCL's can be attributed to the introduction of PCL chains to the OH group of lignins. The change of MR accords well with the results obtained above.

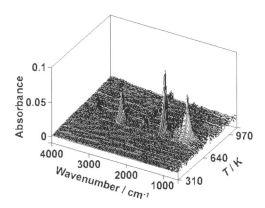

Figure 6-42. Stacked three-dimensional TG-FTIR diagram for KLPCL (CL/OH ratio = 10mol mol^{-1}) [2].

Figure 6-42 shows a representative TG-FTIR stacked three-dimensional diagram showing the change of IR intensity and wavenumber with temperature in TG-FTIR of KLPCL with CL/OH ratio 10 mol mol^{-1}. The assignment of the main peaks observed for the sample are almost the same as described for CAPCL and CellPCL: 1160 cm^{-1} (vC-O-), 1260 cm^{-1} (-C(=O)-O-C-), 1517 and 1617 cm^{-1} (vC=C), 1770 cm^{-1} (vC=O), 2345 cm^{-1} (vCO$_2$), 2945 cm^{-1} (vC-H) and 3700 cm^{-1} (vOH).

The changes of C-O-C, C=O, CO$_2$, CH, and OH peak intensities with CH/OH ratios in KLPCL's at 653 and 693 K are shown in Figures 6-43 and 6-44. Figure 6-43 shows that the change of the peak intensities corresponding to the above IR absorption peaks at 653 K decrease slightly with increasing CL/OH ratios, suggesting that the evolved gases are not prominently formed at this temperature range. The intensities of the C-O-C, C=O, CH and OH absorption peaks increase recognizably with increasing CL/OH ratios, as shown in Figure 6-44, at 693 K. This result suggests that KLPCL degrades considerably and that the evolved gases with the above groups are formed by the thermal degradation of PCL chains, since the intensities of the C-O-C, C=O and CH absorption peaks, which correspond to the structure of PCL chains, increase markedly with increasing CL/OH ratios.

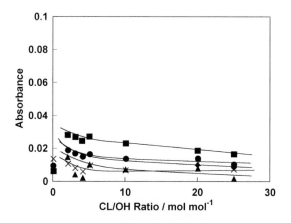

Figure 6-43. Change of C-O-C, C=O, CO_2, CH, and OH peak intensities with CH/OH ratios in KLPCL's at 653 K [2]. ●: C-O-C, ○: C=O, □: CO_2 , △: CH, ▲:OH.

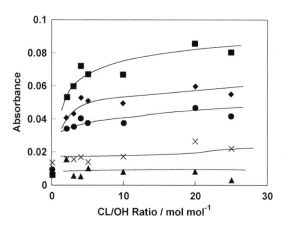

Figure 6-44. Change of C-O-C, C=O, CO_2, CH, and OH peak intensities with CH/OH ratios in KLPCL's at 693 K [2].

The changes of C-O-C and C=O peak intensities with various CH/OH ratios in KLPCL at 653, 673, 693, 703 and 713 K are shown in Figures 6.45 and 6-46. The peak intensities of C-O-C and C=O peaks at 653 and 673 K decrease with increasing CL/OH ratios. This suggests that the evolved gases are not formed by the degradation of PCL chains, but mainly by the degradation of lignin in KLPCL. The C-O-C and C=O peak intensities increase recognizably in the temperature range over 693 K. This peak intensity also increases prominently with increasing CL/OH ratio. The above results indicate that the evolved gas with C-O-C and C=O groups is formed by the thermal degradation of PCL chains. The intensity of the C-O-C and C=O peaks at 703 K is larger than that at 713 K. This may indicate that the evolution of gases caused by PCL chains occurs most prominently in the temperature range around 707 K.

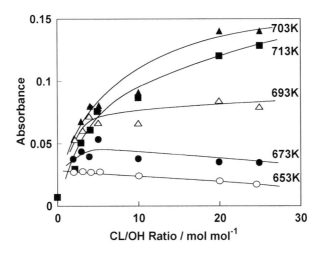

Figure 6-45. Change of C=O peak intensity with CH/OH ratio in KLPCL at various temperatures [2]. ●: 653 K, ○: 673 K, □: 693 K, △:703 K, ▲:713 K.

As described above, environmentally compatible polymers such as polycaproplactone (PCL) derivatives were synthesized from saccharides such as glucose, fructose and sucrose, polysaccharides such as cellulose and cellulose acetate, and lignins such as alcoholysis lignin, kraft lignin and sodium lignosulfonate. Polymerization was initiated by the OH group of saccharides, polysaccharides and lignins.

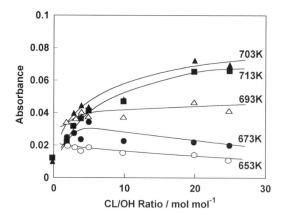

Figure 6-46. Change of CH peak intensity with CH/OH ratio in KLPCL at various temperatures. ●: 653 K, ○ :673 K, □: 693 K, △:703 K, ▲:713 K.

Saccharide- and lignin- based PCL derivatives with various PCL chain lengths were prepared. Glass transition temperatures (T_g's), cold-crystallization temperatures (T_{cc}'s) and melting temperatures (T_m's) of saccharide- and lignin-based PCL's were determined by DSC, and phase diagrams were obtained. T_g's decreased with increasing CL/OH ratio, suggesting that PCL chains act as a soft segment in the amorphous region of PCL derivatives. TG-FTIR analysis of PCL's suggested that compounds having C-O-C, C=O and C-H groups are mainly produced by thermal degradation of PCL chains in saccharide-, cellulose- and lignin-based PCL's. Saccharides, cellulose and lignins efficiently act as hard segments in the above polymers.

REFERENCES

1. Hatakeyama, H., Hirose, S., Teruya, T., Kobashigawa, T. and Tokashiki, T., 2000, Jpat. 3090592.
2. Hatakeyama, H., 2002, Thermal analysis of environmentally compatible polymers containing plant components in the main chain, *J. Therm. Anal. Cal.*, **70**, 755-795.
3. Hatakeyama, H. and Hirose, S., 2002, JPat 3291523.
4. K. Kamide and Sato, M., 1985, Thermal analysis of cellulose acetate solids with total degree of 0.49, 1.75, 2.46 and 2.92, *Polym. J.*, **17**, 919-928.
5. C. G. Pitt, 1990, Bioderadable *Polymers and Drug Delivery Systems* (M. Chasin and R. Langer, eds), Marcel Dekker Inc. New York, pp. 81.
6. Hatakeyama, T., Nakamura, K. and Hatakeyama, H., 1982, Studies on heat capacity of cellulose and lignin by differential scanning calorimetry, *Polymer*, **23**, 1801-1804.
7. Hatakeyama, T. and Liu, Z., 1998, *Handbook of Thermal Analysis*, John Wiley, Chichester (1998) pp. 206.

8. Hatakeyama, H., Yoshida T. and Hatakeyama, T., 2000, The effect of side chain association on thermal and viscoelastic properties of cellulose acetate based polycaprolactones *J. Therm. Anal. Cal.*, **59,** 157-168.

9. Hatakeyama, T. and Quinn, F. X., 1994, *Thermal Analysis*, John Wiley and Sons, Chichester, pp.81-87.

10. Hirose, S. and Hatakeyama, H., 1986, A kinetic study on lignin pyrolysis using the integral method, *Mokuzai Gakkaishi*, **32**, 621-625.

Chapter 7

ENVIRONMENTALLLY COMPATIBLE POLYURETHANES DERIVED FROM SACCHARIDES, POLYSACCHARIDES AND LIGNIN

1. POLYURETHANE DERIVATIVES FROM SACCHARIDES

It is generally recognized that polyurethane (PU) is one of the most useful three-dimensional polymers, since PU has unique features: for example, various forms of materials such as sheets, foams, adhesives and paints can be obtained from PU, and their physical properties can easily be controlled. Over the past 50 years, a number of polyurethanes derived from various polyhydroxyl ingredients and polyisocyanates have been developed in the field of plastics [1]. Plant components having more than two hydroxyl groups per molecule can in principle be used as polyols for PU preparation. In this chapter, new types of polyurethanes derived from mono- and disaccharides (glucose, fructose and sucrose), and molasses are described [2-12].

1.1 Saccharide-based PU sheets

It has been recognized that the plant components act as hard segments in the above PU's and that the thermal and mechanical properties can be controlled in a wide range by changing the amounts of hard and soft segments.

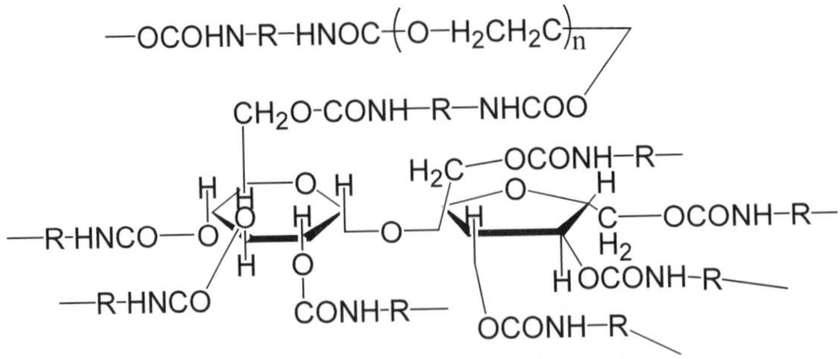

Figure 7-1. Schematic chemical structure of sucrose-based PU. R=core structure of MDI.

The objective of this section is to describe the thermal properties of PU's derived from mono- and disaccharides (glucose, fructose and sucrose). PU sheets were prepared from the saccharide-polyethylene glycol (PEG) – poly(phenylene methylene) polyisocyanate (MDI) system using bulk polymerization [67].

For the preparation of PU's, saccharides such as glucose, fructose and sucrose were first dissolved in PEG 200 (molecular mass 200) or PEG 400 (molecular mass 400) at 323 or 333 K. Prior to reaction with MDI, the polyol solutions of saccharides were dried under vacuum with vigorous stirring at 348 K for 1 hr. Depending on the saccharide content, a 1 % solution of 1, 4-diazabicyclo (2,2,2)-octane (DABCO) in diethylene glycol (DEG) was added to the polyol solution as a catalyst. MDI was added and the reaction was allowed to proceed at room temperature with moderate stirring. The NCO/OH (moles of isocyanate group/ moles of OH groups) ratio was changed from 1.0 to 1.2, depending on the required physical properties of prepared PU sheets. The pre-polymerized mixture was poured into a Teflon coated mold and placed in a hot press at 393 K under a pressure of 10 MPa and subsequently cured in an air-oven between two glass plates. Figure 7-1 shows a schematic chemical structure of sucrose-based PU. The chemical structure of PU is dependent on saccharide component.

1.1.1   Thermal properties of saccharide-based PU sheets

Figure 7-2 shows the change of T_g with the saccharide content. T_g increases steadily with the saccharide content for PU's. The incorporation of saccharides into the PU structure leads to an increase in crosslinking density due to the large number of hydroxyl groups per molecule of the saccharides. The number of hydroxyl groups per molecule of glucose and fructose is 5 mol mol^{-1} and sucrose has a number of hydroxyl groups per molecule of 8 mol mol^{-1}. With the increase of crosslinking density, the main chain motion is more restricted and T_g becomes higher. As well as having a large effect on the crosslinking density, the saccharides act as hard segments that cause an increase in T_g. The MDI content increased with the saccharide content, since the NCO/OH ratio was kept constant. MDI having benzene rings acts as hard segments and thus an increase in the MDI content results in an increment in T_g.

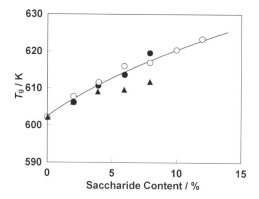

Figure 7-2. Change of glass transition temperatures (T_g's) for PU's containing glucose, fructose and sucrose in the molecular chain measured in N_2. ●: glucose, ○: fructose, ▲: sucrose.

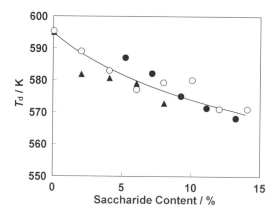

Figure 7-3. Changes of thermal degradation temperatures (T_d's) for PU's containing glucose, fructose and sucrose in the molecular chain measured in N_2. ●: glucose, ○: fructose, ▲: sucrose.

As shown in Figure 7-2, the PU's containing sucrose have lower T_g's than the other samples. Since sucrose contains fewer OH groups per unit of mass than glucose and fructose, the sucrose PU's have lower crosslinking density. At the same time, since the NCO/OH ratio is kept constant, the sucrose PU's have higher PEG content and lower MDI content than the corresponding glucose- and fructose-based PU's. Accordingly, sucrose-based PU's show lower T_g's than glucose- and fructose-based PU's. Figures 7-3 and 7-4 show changes of T_d's of PU's containing glucose, fructose or sucrose in the molecular chain measured in N_2 (Figure 7-3) and in air (Figure 7-4). PU samples containing glucose, fructose and sucrose show similar T_d curves. It can be seen that T_d decreases with increasing saccharide content.

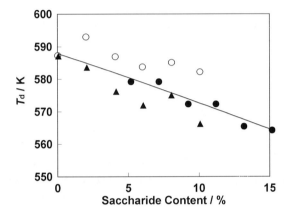

Figure 7-4. Changes of thermal degradation temperatures (T_d's) of PU's containing glucose, fructose and sucrose in the molecular chain measured in air.   ● : glucose, ○: fructose, ▲ : sucrose.

Concerning the above PU's, the relationship between the residue at 773 K and the saccharide content suggested that saccharides constitute a significant part of the residual products. This indicates that the thermal decomposition of the PU's is caused to a fairly large extent by the degradation of PEG and isocyanate portions. The thermal degradation of saccharide portions occurred separately from the degradation of PEG and isocyanates.

1.2   Molasses-based flexible PU foams

Since saccharides are basically biodegradable, PU's with saccharide components are degradable by microorganisms in soil or water [10]. At the same time, it becomes possible to utilize molasses, which is a kind of biowaste as a useful resource for environmentally compatible plastics. Flexible PU foams were prepared from molasses-based polyol (MLP) [6].

The hydroxyl group content of MLP was determined according to JIS K 1557.

Various kinds of polyols for flexible PU foams were prepared by mixing MLP with flexible polyols such as propylene glycol (PPG), graft polyol (GP) and polyester polyol (PEP). As shown at Table 1-2 in Chapter 1, molasses contains sucrose, glucose and fructose as major saccharide components. Several kinds of isocyanates such as toluene diisocyanate (TDI)), lysine diisocyanate (LDI) and lysine triisocyanate (LTI) were used.

1.2.1   Preparation

One type of flexible PU foam was prepared from MLP mixed with polypropylene glycol (PPG 3000, molecular mass 3000) by polymerization with TDI and MDI. Molasses was obtained from Okinawa. Silicon type surfactant, tin (Sn) type catalyst (tin octanoate) and amine catalyst (pentamethyl-diethylenetriamine) were also used for the preparation. The hydroxyl group content of MLP was determined according to JIS K 1557. In order to prepare PU foams, a predetermined amount of PPG 3000 was added to MP. Then calculated amounts of TDI or MDI, surfactant, catalysts and a trace amount of water as a foaming agent were added to MLP and PPG mixture under vigorous stirring. Foaming was carried out immediately after removing the stirrer. The obtained foam was cured overnight at room

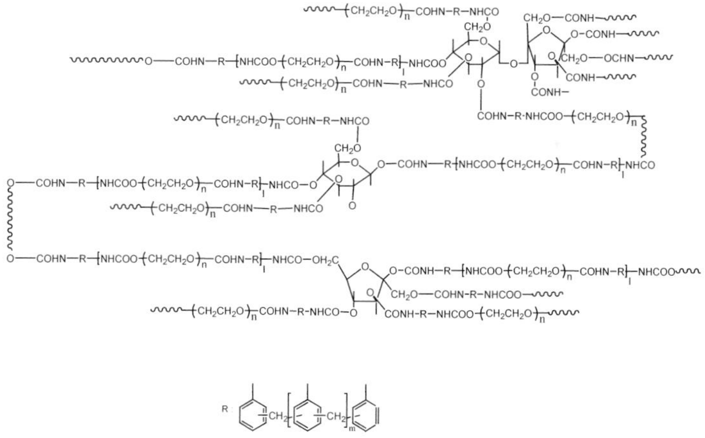

Figure 7-5. Schematic chemical structure of molasses-based flexible PU foams.

temperature [6]. Figure 7-5 shows a schematic chemical structure of the flexible PU foams prepared by the above method.

Another type of flexible PU was prepared according to the preparation scheme shown in Figure 7-6. MLP was first mixed with GP (styrene- and acrylonitrile-grafted polyether) or PEP (molecular weight 2200). Silicon surfactants, catalysts (dibutyltindilaurate, DBTDL, and trimethyl aminoethyl piperazine, TMAEP) and water were added to the solution before mixing. The mixture was reacted with LDI or LTI under vigorous stirring in the presence of dichloromethane. After foams were obtained, the samples were allowed to stand overnight at room temperature. The obtained PU foams were cured at 393 K for 2 hours. The schematic chemical structure of saccharide-based flexible polyurethane foams and chemical structures of raw materials are shown in Figure 7-7.

```
┌─────────────────────────────────────┐
│         Molasses polyol (MLP)        │
│            Graft polyol (GP)         │
│  Polyester polyol (PEP) under heating│
└─────────────────────────────────────┘
                    │
                    │              ┌──────────────┐
                    │◄─────────────│    Water     │
                    │              │  Surfactant  │
                    │              │   Catalyst   │
                    │              └──────────────┘
              ┌──────────┐
              │Premixture│
              └──────────┘
                    │
┌──────────────────┐│  ┌──────────────┐
│ Vigorous stirring│├─►│◄─│ LDI and LTI │
└──────────────────┘│  └──────────────┘
                    │
     ┌──────────────────────────────┐
     │  Flexible polyurethane foams  │
     └──────────────────────────────┘
```

Figure 7-6. Preparation of saccharide-based flexible PU foams [12]. NCO/OH ratio= 1.05.

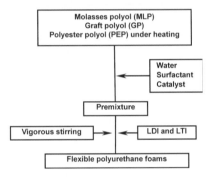

R': GP or PEP

R =

Figure 7-7. Schematic chemical structure of molasses-based flexible PU foams [12]. R = LDI or LTI.

1.2.2 Thermal Properties

Figure 7-8 shows DSC curves of PU's prepared from the PE-GP-MLP-(LDI/LTI) system. Figure 7-9 shows change of T_g with LDI and LTI contents of PU's prepared from the PE-GP-MLP- (LDI/LTI) system. As shown in Figure 7-7, LTI has more reactive sites than LDI. The molecular motion of PU molecules, which were prepared using isocyanates containing poly-reactive sites such as triisocyanate, is more restricted than that of PU molecules prepared by using diisocyanate, because of increased crosslinking density. Accordingly, T_g of PU's prepared by using mixtures of LDI and LTI increased with increasing LTI content.

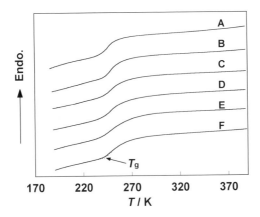

Figure 7-8. DSC heating curves of PU's prepared from the PEP-GP-MLP-(LDI/LTI) systems. Symbols in the figure and LDI / LTI ratios (%) are shown in Table 7-1. Measurements; heat-flux type DSC (Seiko Instruments, DSC 220C), heating rate = 10 K min^{-1}, N$_2$ gas flow rate = 30 ml min^{-1}, samples mass =ca. 5 mg, aluminum open pans were used. Glass transition temperature (T_g) was recognized as an endothermic shift of the baseline in the DSC curve [13, see Figure 2-8 of Chapter 2].

Table 7-1. LDI and LTI ratio (%) in PU's prepared from the PEP-GP-MLP-(LDI / LTI) system

Symbols in Figure	LDI / %	LTI / %
A	100	0
B	80	20
C	60	40
D	40	60
E	20	80
F	0	100

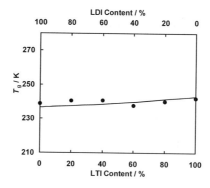

Figure 7-9. Change of glass transition temperatures (T_g's) with LDI and LTI contents in PU's prepared from the PEP-GP-MLP-(LDI/LTI) systems [12].

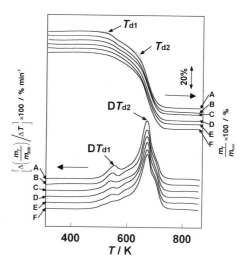

Figure 7-10. TG and DTG curves of PU's prepared from the PEP-GP-MLP-(LDI / LTI) system [12]. Symbols and LDI (%) / LTI (%) ratios are shown in Table 7-1. Measurements; TG-DTA (Seiko Instruments TG/DTA 220) heating rate = 20 K min^{-1}, samples mass = ca. 7 mg, platinum pans were used. N$_2$ flow rate =200 ml min^{-1}. Thermal degradation temperature (T_d) was determined in TG curves according to the method shown in the literature [13, see Figure 2-3 of Chapter 2].

Figure 7-10 shows TG curves and DTG curves of PU's prepared from the PE-GP-MLP-(LDI/LTI) system. Figure 7-11 shows the change of T_d with LDI and LTI contents of PU's prepared from the PE-GP-MLP-(LDI/LTI) system. Two T_d's and DT_d's are observed. The peak of derivative thermal degradation temperature (DT_{d1}) increases with increasing LTI content. This is probably caused by the increase of crosslinking density with increasing LTI components in PU's.

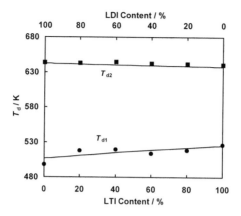

Figure 7-11. Change of thermal degradation temperatures (T_d's) with LDI and LTI contents in PU's prepared from the PEP-GP-MLP-(LDI/LTI) systems [12].

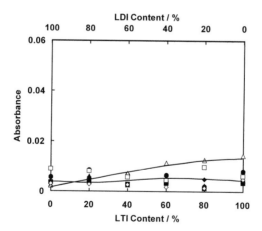

Figure 7-12. IR peak intensities of evolved gases at various wavenumbers plotted against LDI and LTI contents in PU's prepared from the PEP-GP-MLP- (LDI/LTI) system at DT_{d1} (ca. 550 K) [12]. Symbols and absorption bands are shown in Table 7-2. Measurements; TG-Fourier transform infrared spectrometer (TG-FTIR) (Seiko Instruments, TG/DTA220 equipped with Jasco FT/IR-420). heating rate = 20 K min^{-1}, gas flow rate = 100 ml min^{-1}, temperature of the gas transfer system = 540 K, resolution of FTIR = 1 cm^{-1}, one spectrum = 10 scans sec^{-1}.

Table 7-2. IR absorption bands observed by thermal degradation of PE-GP-MLP- (LDI/LTI) systems

Symbols in Figure	Absorption band	Wavenumber / cm^{-1}
●	C-O-C	1134
○	(C=O)-O-C	1206
▲	C=O	1820
△	C=O	1757
■	NCO	2277
□	CO$_2$ and NO$_2$	2363
◆	C-H	2910
◇	H$_2$O	3700

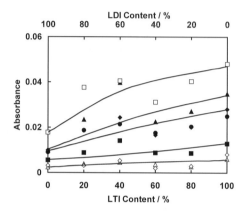

Figure 7-13. IR peak intensities of evolved gases at various wavenumbers plotted against LDI and LTI contents in PU's prepared from the PE-GP-MLP- (LDI/LTI) system at DT_{d2} (ca. 670 K) [12]. Symbols and absorption bands are shown in Table 7-2. Measurements; see Figure 7-12 caption.

The results of TG-FTIR are shown in Figures 7-12 and 7-13. DT_{d1} seems to correspond to the thermal degradation of LTI, since C=O (1820 cm^{-1}) was observed in the evolved gases. The intensity of C=O peak in LTI (1820 cm^{-1}) increases with increasing LTI content. In the thermal degradation, IR peaks corresponding to C-O-C (1134 cm^{-1}), -C(=O)-O-C-(1206 cm^{-1}), CO$_2$ (2362 cm^{-1}) and C-H (2910 cm^{-1}) were observed. DT_{d2} seems to correspond to the thermal degradation of urethane bonding and polyol, since C-O-C (1136 cm^{-1}), -C(=O)-O-C- (1260 cm^{-1}), C=O (1757 cm^{-1}), NCO (2277 cm^{-1}), CO$_2$ (2363 cm^{-1}) and C-H (2948 cm^{-1}) peaks were observed.

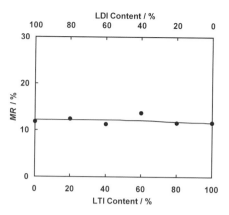

Figure 7-14. Change of mass residueP at 723 K (MR_{723}) with LDI and LTI contents of PU's prepared from the PEP-GP-MLP- (LDI/LTI) system. Mass residue (MR_{723}) was read from TG curve at 723 K [12].

Figure 7-14 shows the change of mass residue at 723 K (MR_{723}) with LDI and LTI contents of PU's prepared from the PEP-GP-MLP-(LDI/LTI) system. MR_{723} does not show obvious change with increasing LTI content. It is considered that LDI and LTI show a similar mass residue, since both LDI and LTI have an aliphatic structure, and accordingly are not heat resistant in the same way as isocyanates having the aromatic structure such as MDI.

1.2.3   Mechanical Properties

Mechanical properties were measured in order to establish the prepared PU foams. Figure 7-15 shows the change of compression strength at 25 % strain (σ_{25}) and compression elasticity (E) of PU's prepared from the PE-GP-MLP-(LDI/LTI) system with LDI and LTI contents. Figure 7-16 shows the change of σ_{25} / ρ and E / ρ of PU's prepared from the PE-GP-MLP-(LDI/LTI) system with LDI and LTI contents. The values of σ_{25}, E, σ_{25} / ρ and E / ρ increase with increasing LTI content, since LTI has more reaction sites than LDI and crosslinking density increases in PU's. The results agree well with the DSC results shown in Figures 7-8 and 7-9.

As described in Section 1.2.1 in this chapter, polyols with long flexible molecular chains such as PPG 3000, PE and GP were introduced into the molecular structure of PU's through the reaction with flexible isocyanates such as TDI, LDI and LTI. Accordingly, PU's having the chemical structures shown in Figures 7-5 and 7-7 could be prepared. These flexible molecular chains gave flexibility to PU foams despite the rigid furanose and pyranose structures of saccharides consisting of molasses.

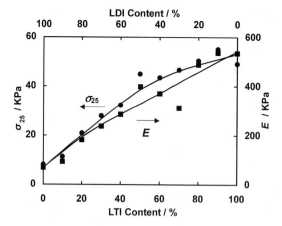

Figure 7-15. Change of compression strength at 25% strain (σ_{25}) and compression elasticity (*E*) with LDI/LTI ratio in PU's prepared from the PEP-GP-MLP-(LDI/LTI) system [12]. Measurements; compression test (JIS K6401 and K7220), mechanical tester, Shimadzu AG-2000D, size of specimen = 40 mm (length) x ca. 40 mm (width) x ca. 30 mm (thickness), temperature = 298 K, Compression speed was 3.0 mm min^{-1}, compression strength (σ_{25}) was detected at 25% strain. The compression elasticity (*E*) was calculated from the gradient of the first straight line of stress-strain curves (JIS K7220).

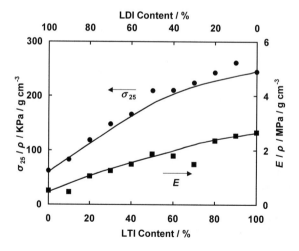

Figure 7-16. Change of compression strength at 25% strain (σ_{25}) / apparent density (ρ) and compression elasticity (*E*) / ρ with LDI/LTI ratio in PU's prepared from the PEP-GP-MLP-(LDI/LTI) system [12].

1.3   Molasses-based semi-rigid PU foams

1.3.1   Preparation

As shown in Figure 7-17, molasses was used as saccharide components of PU's in the preparation of semi-rigid PU foams. Saccharide components give environmental compatibility to PU's through the possibility of biodegradation in soil and water. In order to prepare polyurethane (PU) foams, molasses is first dissolved in PEG, as described in **1.2**. MLP was mixed with polypropylene glycol (PPG, diol type, molecular weight 3000), polyester polyol (PEP, diol type, molecular weight 2500) or polyester polyol (PEP, molecular weight 2200) and small amounts of water, silicone surfactant with the presence of catalysts (di-*n*-butyltin dilaurate, DBTDL, and trimethylaminoethylpiperazine, TMAEP). Polyester polyol was heated from 340 to 348 K when it was used, since polyester polyol has a high viscosity at room temperature.

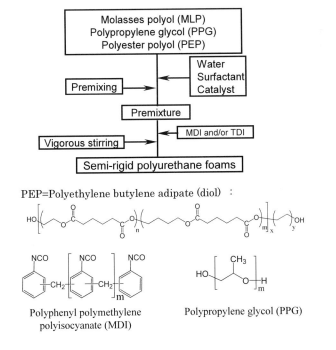

Figure 7-17. Preparation of saccharide-based semi-rigid polyurethane foams [12].

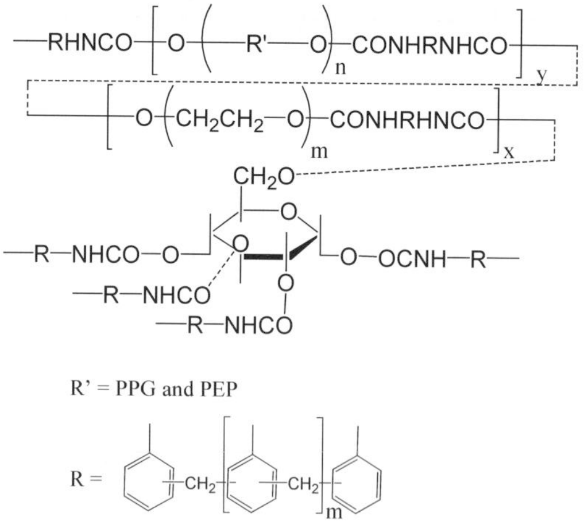

Figure 7-18. Schematic chemical structure of saccharide-based semi-rigid polyurethane foams [12].

The above mixtures were reacted with isocyanates such as poly(phenylene methylene) polyisocyanate (MDI) and/or toluene diisocyanate (TDI) under vigorous stirring. The obtained samples were allowed to stand overnight at room temperature. Schematic chemical structures for PEP, PPG and MDI are shown in Figure 7-17. A schematic chemical structure of saccharide-based PU prepared according to the above methods is shown in Figure 7-18.

1.3.2 Thermal Properties

Figure 7-19 shows the change of T_g with MLP contents in PU's prepared from the PPG-MLP-MDI system. Two T_g's (T_{g1} and T_{g2}) are observed. T_{g1} corresponds to glass transition of PU's with PPG-MLP-MDI domain and T_{g2} corresponds to glass transition of PU's with MLP-MDI domain. It is considered that when MLP content becomes more than 50% phase separation between MLP and PPG occurs. Therefore, two T_g's are observed. It is thought that saccharide acts as a hard segment in the PU, since saccharides in the PU have rigid furanose and pyranose rings having more than 5 reaction sites. The number of sites is much higher than diol and triol type polyols. Accordingly, by using saccharides as components of polyol, crosslinking density of PU's becomes higher than that of standard PU's.

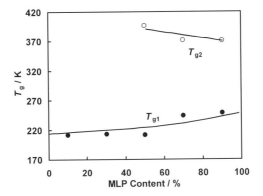

Figure 7-19. Change of glass transition temperatures (T_g's) with MLP contents of PU's prepared from the PPG-MLP-MDI system.

Figure 7-20 shows the change of T_g with PEP contents and PPG contents of PU's prepared from the PEP-PPG-MLP-MDI system. T_g does not show obvious change with the mixing ratio of PEP and PPG. This indicates that because of sufficiently long molecular chains of both polyols (PPG, diol type, molecular weight 3000 and PEP, diol type, molecular weight 2500), the influence on the molecular motion of the prepared PU's is similar.

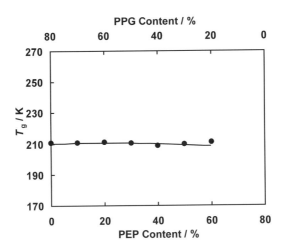

Figure 7-20. Change of glass transition temperatures (T_g's) with PEP and PPG contents of PU's prepared from the PEP-PPG-MLP-MDI system [12].

Figure 7-21 shows DSC curves of PU's prepared from the PEP-PPG-MLP-(MDI/TDI) system. As seen from the figure, T_g is clearly seen as the deviation of the baseline of each DSC heating curve and the shift of T_g with the change of MDI/TDI ratio is also observable.

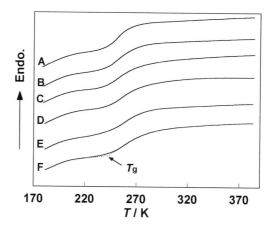

Figure 7-21. DSC heating curves of PU's prepared from the PEP-PPG-MLP-(MDI / TDI) system. MDI (%) / TDI (%) is changed as shown in Table 7-3. Symbols are also shown in Table 7-3 Measurements; heat-flux type DSC (Seiko Instruments DSC 220), sample mass = ca. 5 mg, heating rate = 10 K min^{-1}, N$_2$ gas flow rate = 30 ml min^{-1}. T_g determination; see Figure 2-8 (Chapter 2).

Table 7-3 Symbols and MDI / TDI ratio (%) in PEP-PPG-MLP-(MDI / TDI) systems

Symbols in Figure	MDI / %	TDI / %
A	100	0
B	80	20
C	60	40
D	40	60
E	20	80
F	0	100

Figure 7-22 shows the change of T_g with various MDI and TDI contents, which are shown in Figure 7-21 as the change of MDI/TDI ratio in PU's prepared from the PEP-PPG-MLP-(MDI/TDI) system. T_g increases with increasing MDI content. It is known that MDI is used to prepare rigid polyurethane foams, since MDI has rigid phenyl methane units. The polymeric-MDI has more than two aromatic rings and also has more than two reaction sites (NCO groups). Accordingly, MDI reduces the mobility of the main chain of PU molecules. It is also considered that the increase of MDI content contributes to the increase of the crosslinking density of PU's and that MDI acts as a hard segment in PU molecules.

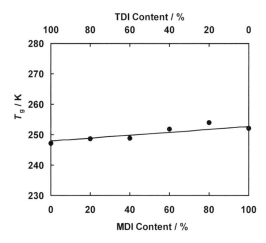

Figure 7-22. Change of glass transition temperatures (T_g's) with MDI and TDI contents of PU's prepared from the PEP-PPG-MLP- (MDI/TDI) system [12].

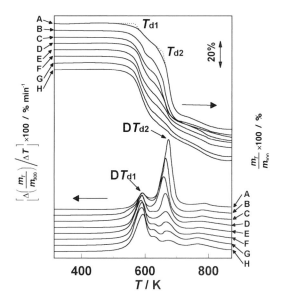

Figure 7-23. TG and DTG curves of PU's prepared from the PPG-MLP-MDI system. MLP (%) / PPG (%) ratio is changed as shown in Table 7-4. Symbols in the figure are also shown in Table 7-4. Measurements; TG-DTA (Seiko Instruments TG/DTA220), samples = ca. 7 mg, platinum pans, heating rate = 20 K min^{-1}, nitrogen gas flow rate = 100 ml min^{-1}. Determination of thermal degradation temperature (T_d); see Figure 2-3 (Chapter 2).

Table 7-4. Symbols and MLP / PPG ratio in PPG-MLP-MDI systems

Symbols in Figure	MLP / %	PPG / %
A	10	90
B	20	80
C	30	70
D	50	50
E	60	40
F	70	30
G	80	20
H	90	10

Figure 7-23 shows TG curves and DTG curves of PU's with various MLP contents in the PPG-MLP-MDI system. Two T_d's (T_{d1} and T_{d2}) are observed as shown in Figure 7-22. As shown in Figure 7-22, the peak of DT_{d1} increased with increasing MLP content and the peak of DT_{d2} increased with increasing PPG content. Accordingly, it is considered that DT_{d1} corresponds to the thermal degradation of MLP and DT_{d2} corresponds to the thermal degradation of PPG. Figure 7-24 shows the change of T_{d1} and T_{d2} with increasing MLP content. The presence of two kinds of thermal degradation is clearly seen.

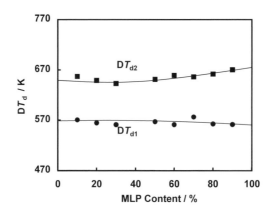

Figure 7-24. Change of derivative thermal degradation temperatures (DT_d's) with MLP contents of PU's prepared from the PPG-MLP-MDI system [12].

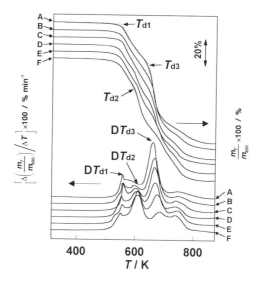

Figure 7-25. TG and DTG curves of PU's prepared from the PEP-PPG-MLP-MDI system. Symbols and PPG / PE / MLP ratio (%) are shown in Table 7-5. Measurements; see Figure 2-24 caption (Chapter 2).

Table 7-5. Symbols and PPG/ PEP / MLP ratio (%) in PEP-PPG-MLP-MDI systems

Symbols in Figure	PPG / %	PEP / %	MLP / %
A	70	10	20
B	60	20	20
C	50	30	20
D	40	40	20
E	30	50	20
F	20	60	20

Figure 7-25 shows TG and DTG curves of PU's with various PEP and PPG contents in the PEP-PPG-MLP-MDI system. Three T_d's (T_{d1}, T_{d2} and T_{d3}) are observed in TG curves in Figure 7-25. The peak of DT_{d1} does not change with mixing ratios of PEP and PPG. The peak of DT_{d2} increases with increasing PEP content and the peak of DT_{d3} decreases with increasing PEP content. Accordingly, it is considered that DT_{d1} corresponds to the thermal degradation of MLP, DT_{d2} corresponds to the thermal degradation of PEP and DT_{d3} corresponds to the thermal degradation of PPG. As shown in Figure 7-26, the presence of three kinds of DT_d is clearly seen and this indicates that three kinds of thermal degradation occur separately with the change of temperature.

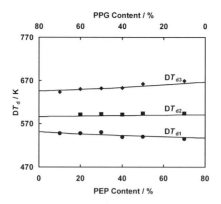

Figure 7-26. Change of derivative thermal degradation temperatures (DT_d's) with PEP, PPG and MLP contents of PU's prepared from the PEP-PPG-MLP-MDI system [12].

Figure 7-27 shows TG and DTG curves of PU's with various MDI and TDI contents in PU's prepared according to the PEP-PPG-MLP-(MDI/TDI) system. Two-step degradation is observed. Derivative thermal degradation at ca. 570 K (DT_{d1}) is smaller than DT_{d2}, which is observed at ca. 650 K, when TDI content is small, but becomes prominent with the increase of TDI content. On the other hand, DT_{d2} peak is prominent when MDI content is high. Accordingly, it is considered that DT_{d1} corresponds to the degradation of TDI component and DT_{d2} corresponds to that of MDI. Figure 7-28 shows the change of DT_{d1} and DT_{d2} with the change of MDI/TDI ratio. The above results indicate that thermal degradation of PU components occurs separately in PU's and this degradation is dependent on the characteristics of each component

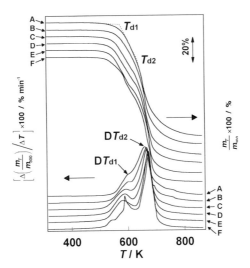

Figure 7-27. TG and DTG curves of PU's prepared from the PEP-PPG-MLP-(MDI/TDI) system. MDI (%) / TDI (%) ratio is changed as shown in Table 7-3. Measurements; see Figure 2-24 caption (Chapter 2).

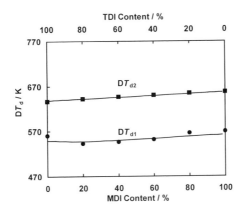

Figure 7-28. Change of derivative thermal degradation temperatures (DT_d's) with TDI contents and MDI contents of PU's prepared from the PEP-PPG-MLP-(TDI/MDI) system.

Figure 7-29 shows the change of MR_{723} with MLP contents of PU's prepared from the PPG-MLP-MDI system. MR_{723} increases with increasing MLP content. MDI contents in PU's increased with the increase of MLP content in polyol, since MLP has hydroxyl value (OHV, OHV_{MLP}=9.87 m mol g^{-1}) which is much higher than that of PPG (OHV_{PPG} = 0.66 m mol g^{-1}). It is known that the fragments from aromatic rings in polymer chains remain until high temperature regions in the residues [9]. Therefore, it is considered that the increase of MR_{723} is most probably caused by the increase of aromatic rings coming from the MDI component in PU's.

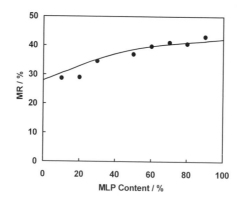

Figure 7-29. Change of mass residue at 723 K (MR_{723}) with MLP contents of PU's prepared from the PPG-MLP-MDI system [12].

Figure 7-30 shows the change of MR_{723} with PPG and PEP contents in PU's prepared from the PEP-PPG-MP-MDI system. MR_{723} values do not change with the change of PPG and PEP contents. This indicates that PPG and PPG degrade at the same rate even if the degradation temperature is different, as shown in Figure 7-26, since both polymers are those with aliphatic structures.

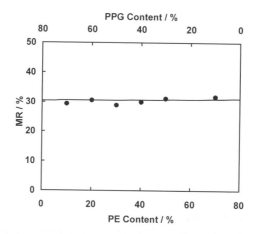

Figure 7-30. Change of mass residue at 723 K (MR_{723}) with PEP and PPG contents in PU's prepared from the PEP-PPG-MLP-MDI system [12].

Figure 7-31 shows the change of MR_{723} with MDI and TDI contents in PU's prepared from the PEP-PPG-MLP-(MDI/TDI) system. MR_{723} increases with increasing MDI content. This is caused by the fact that the increase of MDI contents in PU's increases the amount of aromatic rings in the PU structure. Since MDI has more aromatic structures than TDI, the increase of MR_{723} is mainly caused by the increase of aromatic ring structures in PU's with increasing MDI content.

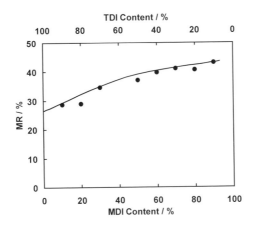

Figure 7-31. Change of mass residue at 723 K (MR_{723}) with MDI and TDI contents of PU's prepared from the PEP-PPG-MLP-(MDI/TDI) system [12].

1.3.3   Mechanical Properties

Figures 7-32 and 7-33 show the change of σ_{10}, E, σ_{10}/ρ and E/ρ values of PU's prepared from the PEP-PPG-MLP-MDI system. The above values increase with increasing PEP content until the content reaches 60 % and then decrease with PEP content over 60 %. This indicates that mechanical strength of PEP 2500 component in PU's is higher than that of PPG. The sudden decrease of the σ_{10} and E values in the PEP content over 60 % seems to be caused by the phase separation of PPG and PEP because of the difficulty of making homogeneous PEP-PPG solution over this PEP content.

Figures 7-34 and 7-35 show the change of σ_{10}, E, σ_{10}/ρ and E/ρ values of PU's prepared from the PEP-PPG-MLP-(TDI/MDI) system. The above values increase with increasing MDI content, indicating that the MDI component in PU's is mechanically stronger than that of TDI.

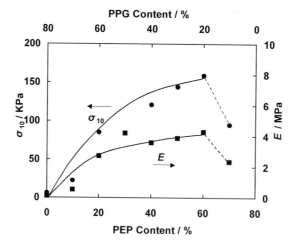

Figure 7-32. Change of compression strength at 10% strain (σ_{10}) and compression elasticity (E) with PEP and PPG contents of PU's prepared from the PEP-PPG-MLP-MDI system [12].

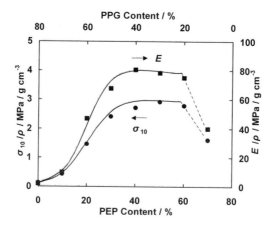

Figure 7-33. Change of compression strength at 10 % strain (σ_{10})/ apparent density (ρ) and compression elasticity (E) / ρ with PEP and PPG contents in PU's prepared from the PEP-PPG-MLP-MDI system [12].

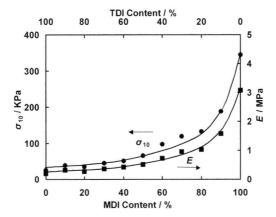

Figure 7-34. Change of compression strength at 10% strain (σ_{10}) and compression elasticity (E) with TDI and MDI contents in PU's prepared from the PEP-PPG-MLP-(TDI/MDI) system [12].

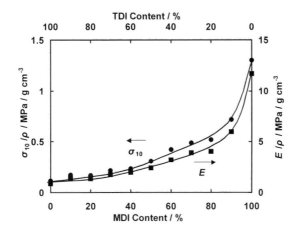

Figure 7-35. Change of compression strength at 10 % strain (σ_{10}) / apparent density (ρ) and compression elasticity (E) / ρ with TDI and MDI contents in PU's prepared from the PEP-PPG-MLP-(TDI/MDI) system [12].

2.   POLYURETHANES DERIVED FROM LIGNINS

As raw materials for the preparation of lignin-based PU's, various kinds of industrial lignins can be used. The following methods are examples of preparation of lignin-based PU's [7]. Prior to obtaining PU's, kraft lignin (KL), alcoholysis lignin (AL), solvolysis lignin (SL), which was obtained as a by-product in organosolve pulping of Japanese beech (*Fagus crenata*) with aqueous cresol at 185 °C without an acid catalyst, and sodium lignosulfonate (LS) were dissolved in polyols such as PEG and PPG in order to prepare polyol solutions containing lignin. The obtained polyol solutions were mixed with MDI at room temperature, and precured polyurethanes were prepared. Each of the precured polyurethanes was heat-pressed and a PU sheet was prepared. In order to prepare PU foams, one of the above lignin-based polyol solutions was mixed with a plasticizer, surfactant (silicon oil), and a catalyst (DBTDL), and then MDI was added. This mixture was stirred with a trace amount of water which was added as a foaming agent. In the above processes, the NCO/OH ratio, the weight of starting materials and the contents of lignin (shown as Lig in the following equations), the amount of polyols such as PEG and PPG (shown as PEG in the following equations), and MDI were calculated according to the following equations:

$$NCO/OH = M_{MDI} + W_{MDI}\left(M_{Lig} x W_{Lig} + M_{PEG} + W_{PEG}\right) \qquad (7.1)$$

$$W_t = W_{Lig} + W_{PEG} \qquad (7.2)$$

$$\text{Lignin content (\%)} = \left(\frac{W_{Lig}}{W_t}\right) x100 \qquad (7.3)$$

$$\text{PEG content (\%)} = \left(\frac{W_{PEG}}{W_t}\right) x100 \qquad (7.4)$$

where M_{MDI} is the number of moles of isocyanate groups per gram of MDI, W_{MDI} the weight of MDI, M_{Lig} the number of moles of hydroxyl groups per gram of lignin, W_{Lig} the weight of lignin, M_{PEG} the number of moles of hydroxyl groups per gram of PEG, W_{PEG} the weight of PEG, and where W_t is the total weight of lignin and PEG in the PU system. In some cases, DEG was used instead of PEG.

2.1 Rigid polyurethane foams derived from kraft lignin

2.1.1 Preparation

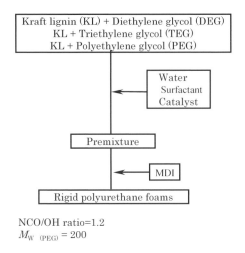

Figure 7-36. Preparation of kraft lignin-based rigid polyurethane foams and chemical structures of DEG, TEG, PEG200 and MDI [12].

In order to prepare PU foams derived from KL (KLPU foams), KL was dissolved in diethylene glycol (DEG), triethylene glycol (TEG) or PEG200 under heating from 338 to 348 K. As shown in Figure 7-36, the above solutions with various KL contents from 0 to 33 % were mixed with small amounts of silicon surfactant, di-*n*-butyltin dilaurate (DBTDL) and water. This premixture was reacted with MDI under stirring at room temperature. NCO/OH ratio was 1.2. PU foams obtained from the above three kinds of lignin-based oligo- and poly-ethyene glycols (DEG, TEG and PEG) are designated as KLDPU, KLTPU and KLPPU, respectively. Figure 7-36 shows chemical structures of DEG, TEG, PEG 200 and MDI. Figure 7-37 shows the schematic chemical structure of KL-based PU.

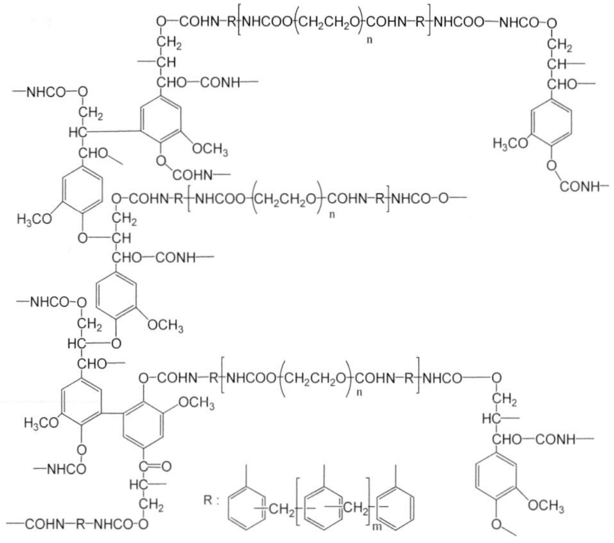

Figure 7-37. Schematic chemical structure of KL-based PU [12].

2.1.2   Thermal properties

Figures 7-38 show DSC curves of KLDPU, KLTPU and KLPPU with various KL contents. Figure 7-39 shows change of T_g with KL contents in KLDPU, KLTPU and KLPPU. T_g's of KLDPU and KLTPU do not change markedly with increasing KL content. This indicates that the molecular chains of DEG and TEG components in the above PU foams, which are short and rigid, restrict efficiently the motion of the above PU foams, and for the above reason, the influence of rigid lignin structure on the molecular motion of KLDPU and KLTPU is relatively small. However, T_g of KLPPU increases with increasing KL content. In the case of KLPPU, KL structure acts as a hard segment in PU network, since PEG component in the PU is relatively long and flexible and the molecular motion is easily restricted by the influence of hard lignin structure.

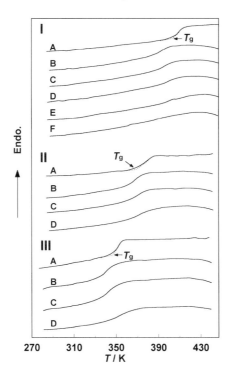

Figure 7-38. DSC curves of KLDPU, KLTPU and KLPPU. KL contents and symbols in the figure are shown in Table 7-6. Measurements; heat-flux type DSC (Seiko Instruments DSC 220), sample mass = ca. 5 mg, heating rate =10 K min⁻¹, N₂ gas flow rate= 30 ml min⁻¹, using aluminium open type pans.

Table 7-6. KL content in KLDPU, KLTPU and KLPPU

Symbols in Figure	Abbreviation	Number of repeating units of ethylene glycol	KL content in polyol / %
I -A	KLDPU	2	0
-B		2	6.6
-C		2	13.2
-D		2	19.8,
-E		2	26.4
-F		2	33.0
II -A	KLTPU	3	0
-B		3	6.6
-C		3	13.2
-D		3	19.8
III-A	KLPPU	4	0
-B		4	6.6
-C		4	13.2
-D		4	19.8

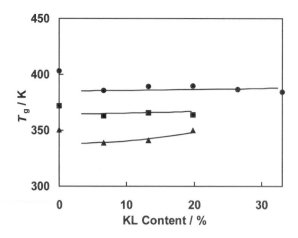

Figure 7-39. Change of glass transition temperatures (Tg's) with KL contents in KLDPU, KLTPU and KLPPU [12]. ●: KLDPU, ■: KLTPU, ▲: KLPPU

Figure 7-40 shows TG and DTG curves of KLDPU with various KL contents. Two step thermal degradations are observed from DTG curves in the above PU's. Figure 7-41 shows change of T_d with KL contents in KLDPU, KLTPU and KLPPU. In KLDPU and KLTPU, T_d does not change obviously with increasing KL content. However, T_d of KLPPU decreases slightly with increasing KL content. Since the increase of KL content in KLPPU reduces the amount of PEG 200, the dissociation between the phenolic hydroxyl group in KL and the isocyanate group becomes prominent, and accordingly T_d decreases. In general, thermal degradation of urethane bonding formed by reaction of phenolic hydroxyl and isocyanate groups occurs at lower temperature than that of urethane bonding between alcoholic hydroxyl groups and isocyanate groups [14, 15].

Change of MR_{723} with KL contents in KLDPU, KLTPU and KLPPU is shown in Figure 7-42. MR_{723} increases in the following order: KLDPU>KLTPU>KLPPU. The hydroxyl values (OHV's) of DEG, TEG and PEG 200 are as follows: 18.9 m mol g^{-1}, 13.2 m mol g^{-1}and 10 m mol g^{-1}, respectively. According to the above data, DEG has the highest OHV contents in the above three glycol's. Therefore, the amount of MDI components in KLDPU is higher than that in KLTPU and KLPPU. It is known that the aromatic ring of MDI in PU polymer chains remains without degradation at high temperature [15]. Therefore, it is considered that the above increase of MR_{723} in KLDPU is mainly caused by the increase in the content of aromatic rings in PU's, since the amount of MDI increases with increasing OHV.

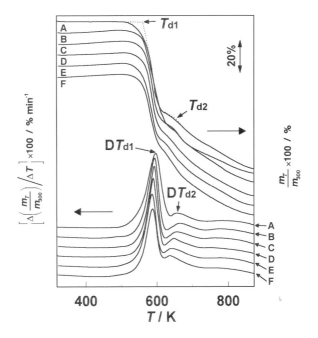

Figure 7-40. TG and DTG curves of KLDPU. KL content in polyol (%): A = 33, B = 26.4, C = 19.8, D = 13.2, E = 6.6, F = 0   Measurements; TG-DTA (Seiko Instruments, TG/DTA 220), sample mass = ca. 7 mg, heating rate = 20 Kmin^{-1}, N$_2$ gas flow rate = 100 ml min^{-1}

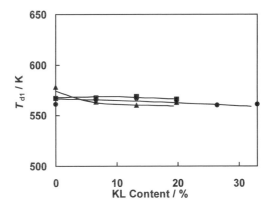

Figure 7-41. Change of thermal degradation temperature (T_{d1}) with KL contents in KLDPU, KLTPU and KLPPU. ●: KLDPU, ■: KLTPU, ▲: KLPPU

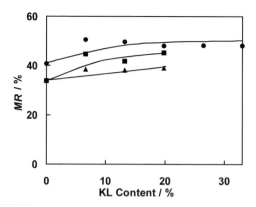

Figure 7-42. Change of mass residue at 723 K (*MR*$_{723}$) with KL contents in KLDPU, KLTPU and KLPPU [12]. ●: KLDPU, ■: KLTPU, н: KLPP.

2.1.3   Mechanical Properties

Figure 7-43 shows stress-strain curves of KLPPU with various KL contents. KLDPU and KLTPU showed similar S-S curves, although σ values are higher than those of KLPPU. As shown in Figure 7-44 (A), *ρ* values decrease with increasing KL content. This is probably caused by the increase of the cell size with increasing KL content. It is known that the increase of

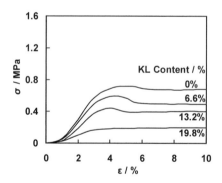

Figure 7-43. Stress-strain curves of KLPPU's with various KL contents [12]. Measurements; compression tests, Shimadzu Autograph AG-2000D (JIS K7220) temperature = 298 K, number of test pieces = 3, size of test piece = ca. 40 mm x ca. 40 mm x ca. 30 mm, compression speed = 3.0 mm min^{-1}. The compression strength (σ_{10}) was detected at 10% strain. The yielding strength (σ_y) was evaluated as the maximum strength from stress-strain curve. The compression elasticity was calculated from the gradient of the first straight line of stress-strain curves according to the JIS K7220.

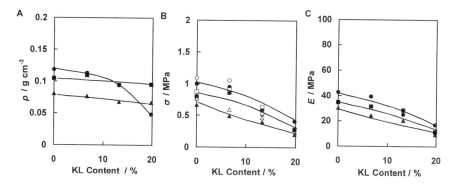

Figure 7-44. Change of apparent density (ρ) (A), compression strength at 10% strain (σ_{10}) and yielding strength. (σ_y) (B), and compression elasticity (*E*) (C) of KLDPU, KLTPU and KLPPU. Symbols are shown in *Table 7-7*.

Table 7-7. Symbols used in Figure 7-44

PU Sample	$\rho*$/ g cm^{-3} Fig. 7-44, A	$\sigma_{10}**$ / MPa Fig. 7-44, B	σ_y***/MPa	$E****$ / MPa Fig.7-44, C
KLDPU	●	●	○	●
KLTPU	■	■	□	■
KLPPU	▲	▲	△	▲

*Apparent density, ** Strength at 10% strain. *** Yielding strength,
**** Compression Elasticity

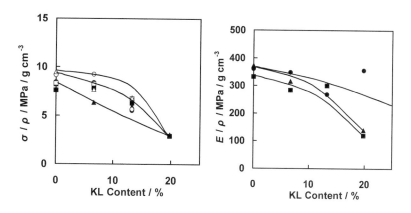

Figure 7-45. Change of compression strength at 10% strain (σ_{10}) / apparent density, (ρ), yielding strength (σ_y) / ρ and compression elasticity (*E*) / apparent density (ρ) with KL contents in KLDPU, KLTPU and KLPPU. Symbols are shown in *Table 7-8*.

Table 7-8. Symblols shown in Figure 7-45

Sample	σ_{10} / ρ: MPa cm^{-3} g^{-1}	σ_y / ρ MPa cm^{-3} g^{-1}	σ_y / E MPa cm^{-3} g^{-1}
KLDPU	●	○	●
KLTPU	■	□	■
KLPPU	▲	△	▲

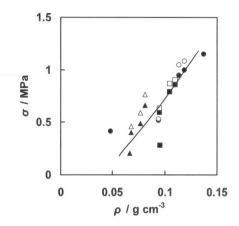

Figure 7-46. Change of compression strength at 10% strain (σ_{10}) and yielding strength (σ_y) with apparent density (ρ) of KLDPU, KLTPU and KLPPU [12]. σ_{10}, ●: KLDPU, ■: KLTPU, ▲: KLPPU, σ_y,○: KLDPU, □: KLTPU, △: KLPPU. Measurements; Apparent density (ρ = weight / apparent volume: g cm^{-3}) was measured using a Mitsutoyo ABS digital solar caliper and an electric balance. Apparent volume was calculated using sample with ca. 40 mm (length)×ca. 40 mm (width)×ca. 30 mm (thickness). Conditions of mechanical tests; see Figure 7-43 caption.

viscosity of KL polyol, which increases with increasing KL content, makes the cell size larger. This decrease of ρ value usually causes the decrease of the compression strength of KLDPU, KLTPU and KLPPU.

Figures 7-44 (B) and 7-44 (C) show the change of σ_{10}, σ_y and E with KL contents of KLDPU, KLTPU and KLPPU. The values of σ_{10}, σ_y and E of KLDPU, KLTPU and KLPPU decrease with increasing KL content. As shown in Figure 7-45, the values of σ_{10}/ ρ, σ_y/ ρ and E/ ρ decrease with increasing KL content. Figures.7-46 and 7-47 show the change of σ_{10}, σ_y and E values of KLDPU, KLTPU and KLPPU with ρ values. The values of σ_{10}, σ_y and E increase linearly with increasing ρ. This indicates that the compression strength and compression elasticity of rigid polyurethane foams highly depend on the values of ρ.

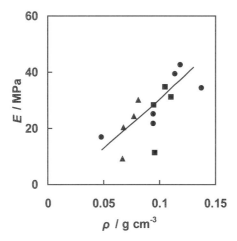

Figure 7-47. Change of compression elasticity (*E*) with apparent density (ρ) of KLDPU, KLTPU and KLPPU [12]. ●: KLDPU, ■: KLTPU, ▲: KLPPU.

The above dependency of mechanical properties of KLPU foams on the ρ values strongly indicates that the morphological properties of foams such as cell size and thickness of cell wall affect the strength and elasticity of foams.

2.2 Rigid polyurethane foams derived from sodium lignosulfonate

Sodium lignosulfonate (LS) was obtained as a by-product of sulfite pulping process. LS has not been used as a chemical component of standard lignin-based polymers such as epoxy resins due to its ionic nature. LS is soluble in water and, at the same time, it is soluble in some kinds of solvent such as ethylene glycol, DEG, TEG and PEG, since LS has an amphiphilic character. It has recently been found that LS-based PU foams (LSPU) can be prepared using DEG, TEG and PEG [16]. In this section, the thermal and mechanical properties of rigid LSPU foams, which were investigated by differential scanning calorimetry (DSC), thermogravimetry (TG), TG-Fourier transform infrared spectroscopy (TG-FTIR) and compression tests, are described.

2.2.1 Preparation

In order to prepare PU foams, LS was dissolved in DEG, TEG or PEG 200 under heating from 338 to 348 K. As shown in Figure 7-48, the above solutions with various LS contents from 0 % to 33 % were mixed with small amounts of silicon surfactant, di-*n*-butyltin dilaurate (DBTDL) and water.

This premixture was reacted with MDI under stirring at room temperature. NCO/OH ratio was 1.2 [16]. PU foams obtained from the above three kinds of lignin-based oligo- and poly-ethylene glycols (DEG, TEG and PEG) are designated as LSDPU, LSTPU and LSPPU, respectively. Figure 7-49 shows a schematic chemical structure of LS-based PU's.

Sodium lignosulfonate (LS) + Diethylene glycol (DEG)
LS + Triethylene glycol (TEG)
LS + Polyethylene glycol (PEG)

Water
Surfactant
Catalyst

Premixture

MDI

Rigid polyurethane foams

NCO/OH ratio=1.2
$M_{W\ (PEG)} = 200$

Figure 7-48. Preparation of sodium lignosulfonate (LS)-based rigid polyurethane foams [16].

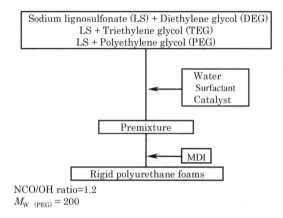

Figure 7-49. Schematic chemical structure of LS-based PU [12].

2.2.2   Thermal properties

Figures 7-50 and 7-51 show DSC curves of LSDPU, LSTPU and LSPPU with various LS contents. In each DSC curve, a gap in the baseline due to glass transition is observed. Figure 7-52 shows the change of T_g with LS contents in LSDPU, LSTPU and LSPPU. T_g's of LSDPU and LSTPU increase slightly with increasing LS content. However, T_g of LSPPU increases clearly from 350 to 364 K with increasing LS content, since the rigid phenyl propane structure in LS acts as a hard segment efficiently in PU networks containing long oxyethylene chains of PEG such as in the case of LSPPU. On the other hand, it is considered that LS molecules in LSDPU and

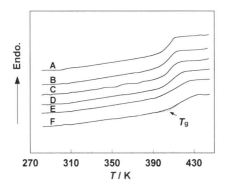

Figure 7-50. DSC heating curves with various LS contents in LSDPU. LS contents in polyol (%): A = 0, B = 6.6, C = 13.2, D = 19.8, E = 26.4, F =33.0   Measurements; see Figure 7-38 caption.

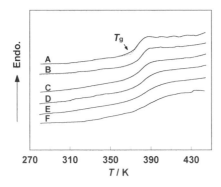

Figure 7-51. DSC heating curves with various LS contents in LSTPU. LS contents in polyol (%): A = 0, B = 6.6, C = 13.2, D = 19.8, E = 26.4, F =33.0.

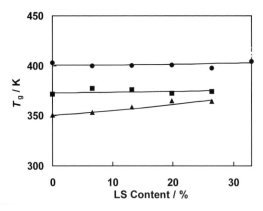

Figure 7-52. Change of glass transition temperatures (T_g's) with LS contents in LSDPU, LSTPU and LSPPU. ●: LSDPU, ■: LSTPU, ▲: LSPPU.

LSTPU do not effectively influence the motion of the PU molecules, since DEG and TEG components in LSDPU and LSTPU have shorter oxyethylene molecular chains than PEG components in LSPPU, which restrict effectively the motion of the PU network in the same way as LS components.

Figure 7-53 shows TG and DTG curves of LSDPU. Two step thermal degradations are clearly observed in DTG curves of the above PU. TG and DTG curves of LSTPU and LSPPU showed a similar tendency. Figures 7-54 and 7-55 show the change of thermal degradation temperatures (T_d's) and derivative thermal degradation temperatures (DT_{d1} and DT_{d2}) with LS contents in LSDPU, LSTPU and LSPPU. In the above PU's, T_{d1} did not change markedly with increasing LS content. However, T_{d2} of LSDPU and LSTPU clearly decrease with increasing LS content. T_{d2} and DT_{d2} of LSDPU, LSTPU and LSPPU decrease noticeably with increasing LS content. It is known that the dissociation of urethane bonding between the phenolic hydroxyl group and the isocyanate group occurs at around 520 K, which is lower than that of urethane bonding between the alcoholic hydroxyl group and the isocyanate group [14, 15]. The increase of LS content in the above LSPU's reduces the amount of polyols such as DEG, TEG and PEG, which are thermally stable compared with LS, when they form urethane bonding with the reaction of isocyanate. Accordingly, T_{d2} and DT_{d2} of LSPU's decrease with increasing LS content in PU's.

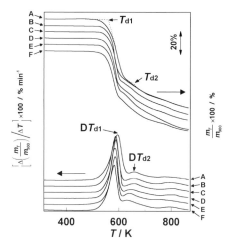

Figure 7-53. TG and DTG curves with various LS contents in LSDPU. LS contents in polyol
(%): A = 33.0, B = 26.4, C = 19.8, D = 13.2, E = 6.6, F =0.

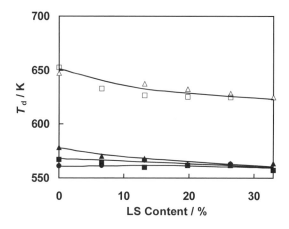

Figure 7-54. Change of thermal degradation temperatures (T_d's) with LS contents in LSDPU,
LSTPU and LSPPU [12]. LSDPU, ▲: T_{d1}, △: T_{d2}; LSTPU, ■: T_{d1}, □: T_{d2}; LSPPU, ●: T_d.

Figure 7-56 shows the change of MR_{723} in LSPU's with LS content in
DEG, TEG and PEG. The results suggest that the MR at 723 K increases
with increasing LS contents, showing that LS contributes to the
improvement of thermal durability of LSPU's.

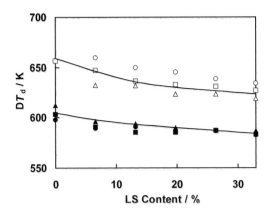

Figure 7-55. Change of derivative thermal degradation temperatures (DT_d's) with LS contents in LSDPU, LSTPU and LSPPU [12]. LSDPU, ●:DT_{d1}, ○:DT_{d2}, LSTPU, ■:DT_{d1}, □:DT_{d2}; LSPPU, ▲:DT_{d1}, △:DT_{d2}.

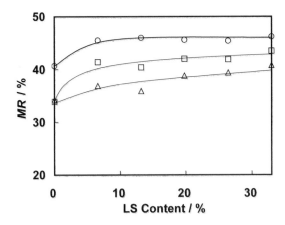

Figure 7-56. Change of mass residue at 723 K (MR_{723}) in LSDPU, LSTPU and LSPPU with LS contents in DEG, TEG and PEG. ○:LSDPU, △: LSTPU, □: LSPPU.

2.2.3   Mechanical properties

Figure 7-57 shows stress-strain curves of LSDPU, LSTPU and LSPPU with 19.8 % LS content in polyol. With increasing LS content, σ_{10} of LSPPU increases while σ_{10} of LSDPU and LSTPU slightly decreases. Mechanical properties of PU foams are markedly affected by the apparent density (r) of PU's. The apparent density can be controlled by foaming conditions, such as viscosity of the reaction mixture and the amount of foaming reagent. When

LS content increases, viscosity of the reaction mixture increases. Although the foaming process is complex, overall structure of PU foams, such as thickness of wall and size of pore in three dimensional structure of PU foams, can be evaluated by apparent density (ρ) of PU foams. Figure 7-58 shows the relationship between ρ and σ_{10}. It is clearly seen that σ_{10} linearly increases with increasing ρ. A similar relationship is observed in the case of compression modulus (E). The above results clearly indicate that mechanical properties of rigid PU foams are dependent on ρ.

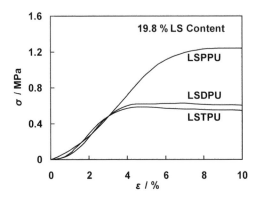

Figure 7-57. Stress-strain curves with 19.8 % LS contents in LSDPU, LSTPU and LSPPU [12], Measurements; see Figure 7-4 caption.

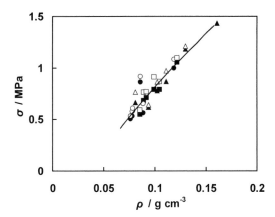

Figure 7-58. Change of compression strength at 10% strain (σ_{10}) and yielding strength (σ_y) with apparent density (ρ) of LSDPU, LSTPU and LSPPU [12]. σ_{10}, ●: LSDPU, ■: LSTPU, ▲: LSPPU; σ_y, ○: LSDPU, □: LSTPU, △: LSPPU.

2.3   Polyurethanes derived from lignin-based PCL's

2.3.1   Preparation

Polyurethanes derived from lignin-based PCL's are obtained according to the following processes. Lignin-based PCL's are first dissolved in tetrahydrofuran (THF). The above solution is reacted with MDI at room temperature with stirring. The obtained prepolymer is cast on a glass plate and the solvent is evacuated under dry conditions. The obtained lignin-based PU sheets are cured at 393 K for 2 hr. The schematic chemical structure of the obtained sample (Lig PCLPU) is shown in Figure 7-59.

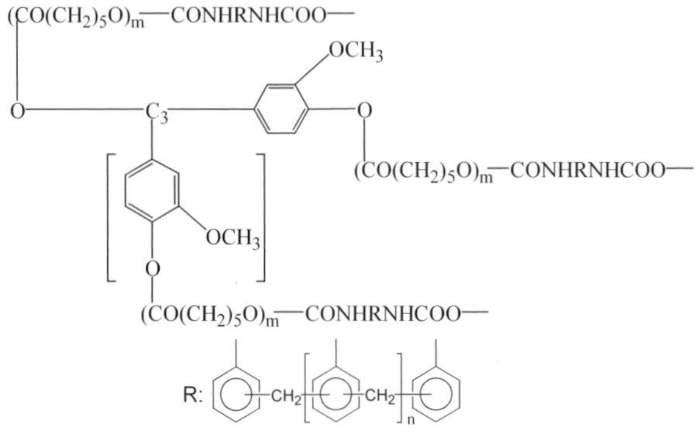

Figure 7-59. Schematic chemical structure of lignin-based PCLPU.

Figure 7-60 shows a part of the DSC curves of PU's derived from alcoholysis lignin-based polycaprolactone (ALPCL) with various CL/OH ratios from 2 to 25 mol mol^{-1}. A marked change in baseline due to glass transition is observed in each DSC curve. T_g decreases with increasing CL/OH ratio from 2 to 10 mol mol^{-1} in PU's from lignin based-PCL's. PCL chains with lignin act as soft segments in PU networks. However, as shown in Figure 7-60, when the CL/OH ratio is 10 to 25 mol mol^{-1}, T_g increases. In the case of the DSC curves representing ALPCLPU with CL/OH ratio 15 mol mol^{-1}, an exothermic peak due to cold-crystallization of ALPCLPU and also a large peak due to melting of crystalline region are observed when CL/OH ratio is over 15. The above results suggest that the PU's derived from ALPCL's with CL/OH ratios over 15 have a clear crystalline region in the molecular structure. A similar phenomenon is observed in PU's from KLPCL's.

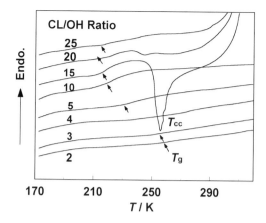

Figure 7-60. DSC heating curves at around glass transition of ALPCLPU.

Figure 7-61 shows the changes of T_g's of PU's from AL- and KL-PCL's. The T_g markedly decreases with increasing CL/OH ratio in the region where CL/OH ratio is below 15 mol mol^{-1} and then the T_g levels off with increasing CL/OH ratio in the region where CL/OH ratio exceeds 15. The leveling off of T_g over CL/OH ratio = 15 mol mol^{-1} suggests that by the introduction of long PCL chains the crystalline region increased and this restricts the motion of PCL chains.

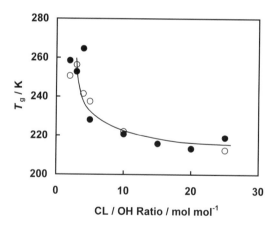

Figure 7-61. Change of T_g with CL/OH ratio of ALPCLPU and KLPCLPU [7]. ○: KLPCLPU ●: ALPCLPU.

Figure 7-62 shows a phase diagram representing changes of T_g's, cold-crystallization temperatures (T_{cc}'s) and melting temperatures (T_m's), against CL/OH ratios of PU's derived from AL- and KLPCL's. The change of T_g's is almost the same for the AL- and KLPCLPU's. T_{cc}'s and T_m's slightly increase with increasing CL/OH ratio in the region of CL/OH ratios over 15, suggesting an increase in the crystalline region of PCL chains in the AL- and KLPCLPU's.

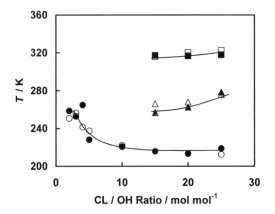

Figure 7-62. Phase diagram of ALPCLPU and KLPCLPU. ALPCLPU [7], ●: T_g, ▲: T_{cc}, ■: T_m ; KLPCL PU, ○: T_g, △: T_{cc}, □:T_m.

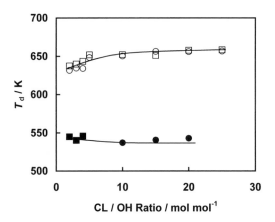

Figure 7-63. Change of T_d with CL/OH ratio of ALPCLPU and KLPCLPU. ALPCLPU, ●: T_{d1}, ○:T_{d2} , KLPCLPU, ■: T_{d1} , □:T_{d2}.

TG and DTG curves of PU's derived from lignin-based PCL's with various CL/OH ratios showed two kinds of thermal degradation temperatures, T_{d1} and T_{d2}. Figure 7-63 shows the change of T_{d1} and T_{d2} with CL/OH ratios of AL- and KL-PCL PU's. It is known that some urethane bonds in PU's dissociate to form hydroxyl and isocyanate groups at about 520 K [14, 15]. Accordingly, it is considered that T_{d1} may reflect the degradation of lignin parts in AL- and KLPCLPU's. As already shown in Figure 6-12 in Chapter 6, T_d of CAPCL increased suddenly and stepped up from 620 to 660 K with increasing CL/OH ratio from that lower than 8 to that higher than 10. The above change of T_d accords similarly with the change of T_{d1} and T_{d2} of AL- and KLPCLPU's as shown in Figure 7-63. Accordingly, it is considered that T_{d1} probably reflects the degradation of the lignin part and T_{d2} may reflect that of PCL parts in AL- and KLPCLPU's [17].

3. SACCHARIDE- AND LIGNIN-BASED HYBRID POLYURETHANE FOAMS

Plant components such as lignin and saccharides are basically biodegradable. They have reactive hydroxyl groups in their molecules and have more than two hydroxyl groups per molecule. Accordingly, it is possible to prepare polyurethane's (PU's) from plant components by the reaction with isocyanates as mentioned in the former sections (See Chapters 7.1 and 7.2). PU's having plant components show excellent thermal and mechanical properties. They are also biodegradable [10]. In Chapters 7.1 and 7.2, it has been stated that saccharides, such as glucose, fructose, sucrose and molasses (ML), and lignins such as kraft lignin (KL) and sodium lignosulfonate (LS), which are obtained as by-products of pulp production processes, can be used as a polyol component for PU's. In this section (Chapter 7.3), we consider PU foams which are prepared from ML, KL and LS and their thermal and mechanical properties in order to establish how the changing ratio of lignin and ML affects the physical properties of PU's.

3.1 Preparation

In order to prepare PU foams, KL and LS are dissolved in PEG 200 under heating from 340 to 350 K. Each of the solutions in the above mixtures is mixed with molasses polyol (MLP) in the presence of small amounts of silicon surfactant, di-*n*-butyltin dilaurate (DBTDL) and water. This premixture is reacted with MDI at room temperature [17]. After foams are prepared, the samples are allowed to stand overnight at room temperature. In the above processes, KL, LS and ML in each solution can be calculated as follows:

$$\text{KL content / \% in polyol} = \left(\frac{W_{KL}}{W_{KL} + W_{ML} + W_{PEG200}} \right) x100 \qquad (7.5)$$

$$\text{LS content / \% in polyol} = \left(\frac{W_{LS}}{W_{LS} + W_{ML} + W_{PEG200}} \right) x100 \qquad (7.6)$$

$$\text{ML content / \% in polyol} = \left(\frac{W_{ML}}{W_{ML} + W_{KLorLS} + W_{PEG200}} \right) x100 \qquad (7.7)$$

where W_{KL} is weight of KL, W_{LS} is weight of LS and W_{ML} is weight of ML.

Figure 7-64 shows preparation scheme of KL-, LS- and ML-based rigid polyurethane foams. Figure 7-65 shows a schematic chemical structure of prepared hybrid type PU.

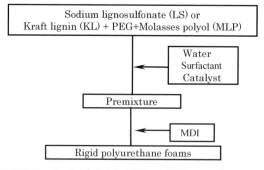

NCO/OH ratio=1.4 (KL-ML-PEG200-MDI system)
NCO/OH ratio=1.2 (LS-ML-PEG200-MDI system)
$M_{W (PEG)} = 200$

Figure 7-64. Preparation of KL- or LS- and ML-based rigid polyurethane foams [17].

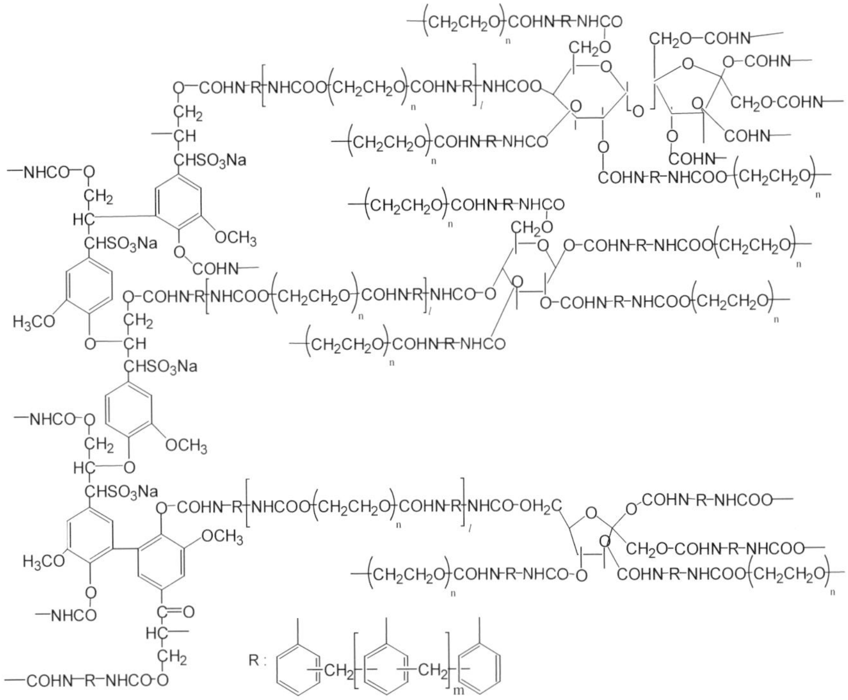

Figure 7-65. Schematic chemical structure of LS- and ML-based polyurethanes [12].

3.2 Thermal properties

Figures 7-66 and 7-67 show stacked DSC curves of PU's with various ML, KL and LS contents, which are prepared from the KL-ML-PEG200-MDI system and the LS-ML-PEG200-MDI system. Figure 7-68 shows change of T_g of PU's with various ML, KL and LS contents. T_g of PU's prepared from the KL-ML-PEG200-MDI system decreases with increasing KL content. However, T_g of PU's prepared from the LS-ML-PEG200-MDI system does not change obviously with increasing LS content. In the preparation of PU's from the KL-ML-PEG200-MDI system and the LS-ML-PEG200-MDI system, an increase of KL and LS contents reduces the amount of MDI, since the increase of KL and LS contents in the above systems causes a decrease in the amount of ML. ML has higher hydroxyl value (OHV) than KL and LS. Saccharides in ML, which have inflexible pyranose and furanose rings, have more reaction sites than lignin. Accordingly, saccharides form more crosslinking sites than lignin in the PU network. Lignin has extremely rigid aromatic rings in the molecular structure, which make the prepared PU rigid. Therefore, the effect of saccharide and lignin structures on T_g of the prepared PU foams creates a balance because of the above factors in the saccharide and lignin structures. T_g of PU's prepared from the KL-ML-PEG200-MDI system is higher than that of PU's prepared from the LS-ML-PEG200-MDI system, since NCO/OH ratio of PU's prepared from the KL-ML-PEG200-MDI system (NCO/OH =1.4) is higher than that of PU's prepared from the LS-ML-PEG200-MDI system (NCO/OH =1.2). Extra amounts of MDI (0.4 for KL-ML-PEG200-MDI system and 0.2 for LS-ML-PEG200-MDI system) react with urethane again, and urea is also formed by the reaction between MDI and water which provides CO_2 gas for blowing. Three-dimensional crosslinking, such as allophanate structure which is formed by the reaction between urethane and isocyanate, and also biuret structure which is formed by the reaction between urea and isocyanate may cause higher T_g of the PU'S from the KL-ML-PEG200-MDI system. Since the content of the above three-dimensional crosslinking structures is higher in PU's prepared from the KL-ML-PEG200-MDI system than PU's prepared from the LS-ML-PEG200-MDI system, it is considered that the mobility of molecular chain of PU's prepared from the KL-ML-PEG200-MDI system is less than that of PU's prepared from the LS-ML-PEG200-MDI system. Therefore, T_g of PU's prepared from the KL-ML-PEG200-MDI system is higher than that of PU's prepared from the LS-ML-PEG200-MDI system.

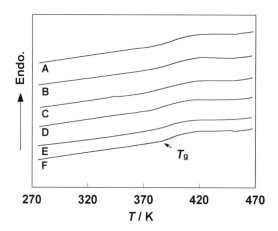

Figure 7-66. DSC heating curves of PU's prepared from the KL-ML-PEG200-MDI system. Symbols in the figure and KL and ML contents in polyol: are shown in Table 7-9. Measurements; heat flux type DSC (Seiko Instruments, DSC 220C), samples mass = ca. 5 mg, aluminum open pans, heating rate =10 K min^{-1}, N_2 gas flow rate= 30 ml min^{-1}. T_g determination; see Figure 2-8 (Chapter 2).

Table 7-9. Composition of kraft lignin and molasses in KL- or LS-ML-PEG200-MDI systems.

Symbols in Figs. 7-66 and 7-67	KL or LS content in polyol/ %	ML content in polyol / %
A	0	16.5
B	3.3	13.2
C	6.6	9.9
D	9.9	6.6
E	13.2	3.3
F	16.5	0

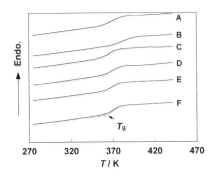

Figure 7-67. DSC curves of PU's prepared from the LS-ML-PEG200-MDI system. Symbols in the figure and LS and ML contents in polyol are shown in Table 7-9. Measurements; see Figure 7-66.

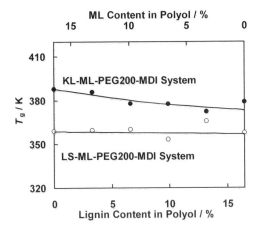

Figure 7-68. Change of glass transition temperature (T_g) with Lignin and ML contents of PU's prepared from the KL-ML-PEG200-MDI system and the LS-ML- PEG200-MDI system [12]. ●: T_g of PU's from the KL-ML-PEG200-MDI system, ○: T_g of PU's from the LS-ML-PEG200-MDI system.

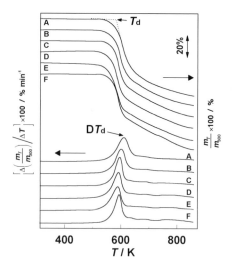

Figure 7-69. TG and DTG curves of PU's prepared from the KL-ML-PEG200-MDI system. Symbols in the figure and KL and ML contents in polyol are shown in Table 7-10. Measurements; TG-DTA (Seiko Instruments, TG/DTA 220), sample mass = ca. 7 mg, heating rate = 20 Kmin^{-1}, N$_2$ gas flow rate = 100 ml min^{-1}.

Figures 7-69 and 7-70 show TG and DTG curves of PU's, which are prepared from the KL-ML-PEG200-MDI system and the LS-ML-PEG200-MDI system. Figure 7-71 shows the change of T_d and MR_{723} with KL and LS contents in PU's from the KL-ML-PEG200-MDI system and the LS-ML-PEG200-MDI system. T_d of PU's does not change obviously with increasing KL and LS contents.

MR_{723} does not show obvious change with increasing KL and LS contents. Since the number of the hydroxyl group in each saccharide unit in saccharides such as glucose, fructose and sucrose in ML is more than that of lignin, the amount of MDI which reacts with the hydroxyl group increases with ML contents. Therefore, this increase of MDI seems to be the reason for the increase of MR_{723} with increasing amounts of saccharides in polyol. On the other hand, lignin is known to be crosslinked by condensation reaction at high temperature. This also causes the increase of MR_{723}. Accordingly, MR_{723} creates a balance in PU's prepared from the KL-ML-PEG200-MDI system and the LS-ML-PEG200-MDI system.

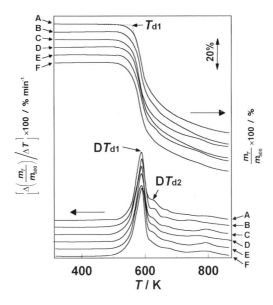

Figure 7-70. TG and DTG curves of PU's prepared from the LS-ML-PEG200-MDI. Symbols in the figure and LS and ML contents in polyol are shown in Table 7-10. Measurements; TG-DTA (Seiko Instruments, TG/DTA 220), sample mass = ca. 7 mg, heating rate = 20 Kmin^{-1}, N$_2$ gas flow rate = 100 ml min^{-1}.

Table 7-10. Composition of kraft lignin and molasses of PU's prepared from the KL- or LS-ML-PEG200-MDI system.

Symbols in Figs. 7-69 and 7-70	KL or LS content in polyol/ %	ML content in polyol / %
A	16.5	0
B	13.2	3.3
C	9.9	6.6
D	6.6	9.9
E	3.3	13.2
F	0	16.5

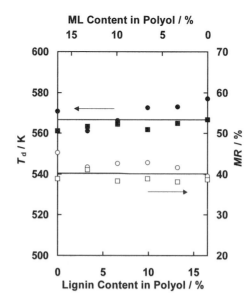

Figure 7-71. Change of thermal degradation temperatures (T_d's) and mass residue at 723 K (MR_{723}) with lignin and ML contents in PU's prepared from the LS-ML- PEG200-MDI system. KL-ML-PEG200-MDI system, ●: T_d, ○: MR_{723}; LS-ML-PEG200-MDI system, ■: T_d, □: MR_{723}.

3.3   Mechanical properties

Figure 7-72 shows stress-strain curves of PU's prepared from the KLP-MLP-PEG200-MDI system and the LS-ML-PEG200-MDI system. Figure 7-73 shows the change of σ_{10}, σ_y and E with various KL and LS contents in PU's from the KLP-MLP-PEG200-MDI system. The values of σ_{10}, σ_y and E increased with KL content. Figure 7-74 shows change of σ_{10}/ρ, σ_y/ρ and E with various KL and LS contents in PU's from the KLP-MLP-PEG200-MDI system. The results are similar to those shown in Figure 7-74.

As shown in Figure 7-75, the values of σ_{10}, σ_y and E increase markedly with the increasing ρ. This result suggests that the mechanical properties depend on ρ. The values of σ_{10}, σ_y and E of PU's prepared from the KL-ML-PEG200-MDI system are larger than those of PU's prepared from the LS-ML-PEG200-MDI system. As shown in Fig. 7-76, these results agree well with T_g values obtained by DSC. The NCO/OH ratio of PU's from the KL-ML-PEG200-MDI system is 1.4 and that of the LS-ML-PEG200-MDI system is 1.2. Therefore, the values of σ_{10}, σ_y and E of PU's prepared from the KL-ML-PEG200-MDI system are higher than those of PU's prepared from the LS-ML-PEG200-MDI system.

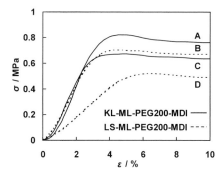

Figure 7-72. The change of σ_{10}, σ_y and E with ρ of PU's prepared from the KL-ML-PEG200-MDI system and the LS-ML-PEG200-MDI system [12]. KL and LS contents in polyol: A = 16.5% KL and 0% ML; B = 16.5% LS and 0 % ML; C = 3.3% KL and 13.2 %ML; D = 3.3% LS and 13.2 % ML.

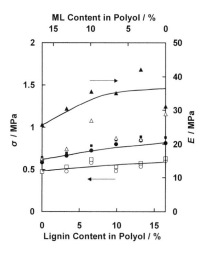

Figure 7-73. Change of compression strength at 10 % strain (σ_{10}), yielding strength (σ_y) and compression elasticity (E) with lignin and ML contents in PU's prepared from the KL-ML-PEG200-MDI system and the LS-ML-PEG200-MDI system [12]. Symbols are shown in Table 7-11

Table 7-11. Symbols used in Figure 7-73

System	Strength at 10 % strain (σ_{10}) /MPa	Yielding strength (σ_y)/ MPa / MPa	Ccompression elasticity (E) / MPa
KL-ML-PEG200-MDI	●	■	▲
LS-ML-PEG200-MDI	○	□	△

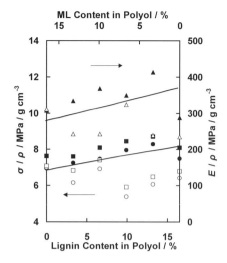

Figure 7-74. Change of σ_{10}/ρ, σ_y/ρ and E/ρ with Lignin contents and ML contents in PU's prepared from the KL-ML-PEG200-MDI system and the LS-ML- PEG200-MDI system [12]. KL-ML-PEG200-MDI system, ●:σ_{10}/ρ, ■:σ_y/ρ, ▲: E/ρ, LS-ML-PEG200-MDI system, ○:σ_{10}/ρ, □:σ_y/ρ, △: E/ρ.

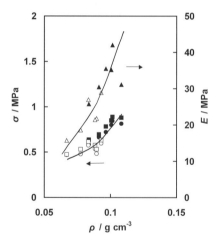

Figure 7-75. Change of compression strength at 10% strain (σ_{10}), yielding strength (σ_y), compression elasticity (E) with apparent density (ρ) of PU's prepared from the KL-ML-PEG200-MDI system and the LS-ML-PEG200-MDI system [12]. KL-ML-PEG200-MDI system, ●: σ_{10}, ■: σ_y, ▲: E ; LS-ML-PEG200-MDI system, ○: σ_{10}, □:σ_y, △: E.

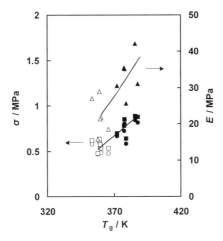

Figure 7-76. Change of compression strength at 10% strain (σ_{10}), yielding strength (σ_y), compression elasticity (E) plotted against T_g of PU's prepared from the KL-ML-PEG200-MDI system and the LS-ML-PEG200-MDI system [12]. KL-ML-PEG200-MDI system, ●: σ_{10}, ■: σ_y, ▲: E; LS-ML-PEG200-MDI system, ○: σ_{10}, □: σ_y, △: E.

From the above results, the following conclusions can be drawn. Rigid polyurethane foams having a variety of thermal and mechanical properties can be derived from polyols containing mixtures of molasses, kraft lignin (KL) or sodium lignosulfonate (LS). T_g's of the PU's prepared from the KL-ML-PEG200-MDI system and the PUs prepared from the LS-ML-PEG200-MDI system do not obviously change with increasing KL and LS contents. T_d's of PU's prepared from the LS-ML-PEG200-MDI system and the KL-ML-PEG200-MDI system do not show clear change with increasing KL and LS content. The value of mass residue at 723 K is different between PU's prepared from the KL-ML-PEG200-MDI system and PU's prepared from the LS-ML-PEG200-MDI system. The values of σ_{10}, σ_y and E of PU's prepared from the KL-ML-PEG200-MDI system and the LS-ML-PEG200-MDI system increase with increasing apparent density.

REFERENCES

1. Szycher, M., 1999, *Szycher's Handbook of Polyurethanes*, CRC Press, Boca Raton, Florida.
2. Hirose, S., Yano. S., Hatakeyama, H. and Nakamura, K. , 2000, *JPat*. 3101701.
3. Hirose, S., Yano. S., Hatakeyama, H. and Nakamura, K. , 1991, *USPat*. 4987213.
4. Hirose, S., Yano. S., Hatakeyama, H. and Nakamura, K. , 1989, *DEPat*. 3884781.
5. Hirose, S., Yano. S., Hatakeyama, H. and Nakamura, K. , 1990, *BRPat*. 8902248.

6. Hatakeyama, H., Hirose, S., Kobashigawa, K.and Tokashiki, T., 1998, JPat. 2836004, *Jpat* 2847497.

7. Hatakeyama, H., 2002, Thermal analysis of environmentally compatible polymers containing plant components in the main chain. *J. Therm. Anal. Cal.* **70**, 755-759.

8. Hatakeyama, H., Asano, Y., and Hatakeyama, T., 2003, Biobased polymeric materials. In *Biodegradable Polymers and Plastics* (Chellini, E. and Solario, R. eds.), Kluwer Academic / Plenum Publishers, New York, pp. 103-119.

9. S. Hirose, K. Kobashigawa and H. Hatakeyama, 1994, Preparation and physical properties of polyurethanes derived from molasses, *Sen-i Gakkaishi*, **50**, 538-542.

10. Morohoshi, N., Hirose S., Hatakeyama, H., Tokashiki, T. and K. Teruya, 1995, Biodegradation of polyurethane foams derived from molasses, *Sen-i Gakkaishi*, **51**, 143-149.

11. Zetterlund, P., Hirose, S., Hatakeyama, T., Hatakeyama, H. and Albertsson, A-C., 1997, Thermal and mechanical properties of polyurethanes derived from mono- and disaccharides, *Polym. Inter.* **42**, 1-8.

12. Asano, Y., 2003, Preparation and physical properties of polyurethane foams derived from saccharides and lignins, Doctoral Thesis of Fukui University of Technology, Fukui, Japan.

13. Hatakeyama, T. and Quinn, F.X. Quinn, 1994, *Thermal Analysis Fundamentals and Applications to Polymer Science*, John Wiley & Sons, Chichester.

14. Hirose, S., Kobashigawa, K., Izuta, Y. and Hatakeyama, H., 1998, Thermal degradation of polyurethanes containing lignin structure by TG-FTIR, *Polymer International*, **47**, 1-8.

15. Saunders, J. H. and Fisch, K., 1962, *Polyurethanes, Chemistry and Technology in High Polymers,* Vol. XV, Interscience Publishers New York.

16. Hirose, S. and Hatakeyama, H., *Jpat.* 2001-223028; *WO* 2002-102873

17. Hatakeyama, H and Hirose, S., 2002, *Jpat.* 3341115..

Chapter 8

BIO- AND GEO-COMPOSITES CONTAINING PLANT MATERIALS

1.   BIOCOMPOSITES CONTAINING CELLULOSE POWDER AND WOOD MEAL

The polyurethanes (PU's) which were prepared from saccharides and lignin showed excellent mechanical and thermal properties [1-10]. They are biodegraded by microorganisms when placed in soil [11]. In this section, composites that are prepared from the above PU's and ground plant particles or powder, such as cellulose powder (CP) and wood meal (WM) are described. Mechanical and thermal properties of the above composites are also cosidered.

1.1   Preparation

PU composites can be prepared according to the scheme shown in Figure 8-1 [7]. As shown in Figure 8-1, cellulose powder or wood meal is mixed with polyols containing molasses. The suspensions with various mixing ratios from 10, 20, 30, 40, 50, 60, 70, 80 and 90 wt % of cellulose powder or wood meal in molasses polyol (MLP) are first prepared [7]. MDI is added to the suspension under stirring and PU composites are prepared. After drying at room temperature, the sample is cured at 393 K for 2 hrs.

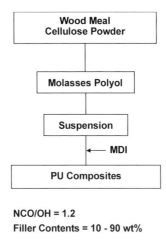

NCO/OH = 1.2
Filler Contents = 10 - 90 wt%

Figure 8-1. Preparation scheme of polyurethane composites (PU composites) [7].

1.2   Thermal and mechanical properties

Figure 8-2 shows change of the density (ρ) of PU composites prepared from cellulose powder and wood meal with the powder content. The density reaches a maximum when the content of plant particles in the composites is from ca.50 % to 70 %.

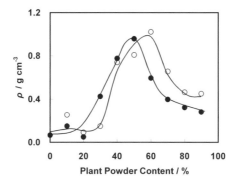

Figure 8-2. Change of the apparent density (ρ / g cm^{-3}) with cellulose powder and wood meal contents in MLP. ●: wood meal ○: cellulose powder. Apparent density (ρ) was measured using a digital solar caliper and an electronic balance. Size of the composite sample was 40-60 mm (length), 20-30 mm (width) and 20-30mm (thickness).

Figure 8-3 shows change of σ with cellulose and wood meal contents in PU composites. As seen from the figure, σ increases with increasing plant powder contents in PU composites, reaches a maximum, and then decreases.

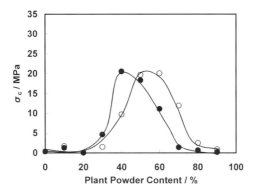

Figure 8-3. Change of compression strength (σ_c) with cellulose powder and wood meal contents in MLP. ●: wood meal, ○: cellulose powder Compression measurements were carried out using a Shimadzu Autograph AG 2000-D at room temperature. Test specimens were a rectangular solid, and the added stress was less than 10 MPa min^{-1}. Compression stress (σ) was defined at the final point of linear compression in the stress-strain curve. Static Young's modulus (E) was calculated using the initial stage of compression curves. Conditions in detail accorded with the Japanese Industrial Standard (JIS Z-2101).

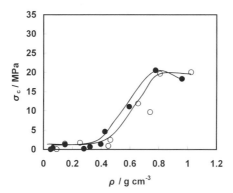

Figure 8-4. Change of compression strength (σ) with density (ρ) of PU composites obtained from cellulose powder and wood meal. ●: wood meal, ○: cellulose powder.

Figure 8-4 shows change of σ with ρ of PU composites obtained from cellulose powder and wood meal. As seen from the figure, σ increases with increasing ρ of PU composites. The above results suggest that the mechanical properties of PU composites from plant powder have a strong relationship with the density of composites: that is to say, the highest mechanical properties are observed when the density of PU composite becomes the highest value.

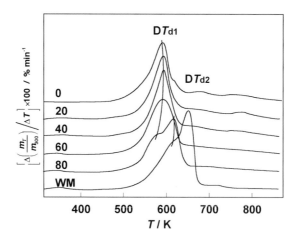

Figure 8-5. DTG curves of wood meal-MLP type PU composites. Measurements; TG-DTA (Seiko Instruments TG/DTA 220), sample mass = ca. 5 mg, heating rate = 20 K min^{-1}, N$_2$ gas flow rate = 100 ml min^{-1}. Mass residue (*MR*) was indicated as $[(m_T - m_{300})/m_{300}]$ x 100, (%), where m_T is mass at temperature T and m_{300} is mass at 300 K. Mass residue was evaluated at 723 K.

Figure 8-5 shows derivative thermal degradation (DTG) curves of PU composites from wood meal. As seen from Figure 8-5, DTG curves show the presence of two kinds of thermal degradation temperatures (T_d's) corresponding to DT_{d1} and DT_{d2}. DT_{d2} seem to be specific to the degradation of wood meal, since the DT_{d2} peak becomes prominent when wood meal contents in PU composites are over 60 % and it is clear when wood meal content is 100 %.

Figure 8-7 shows change of *MR* at 723 K with increasing wood meal content in PU composites, suggesting that wood meal obviously decomposes at 723 K.

As mentioned above, the compression strength (σ), as well as the compression modulus (E), are almost constant in the region of plant powder content lower than 50 %. When the plant powder content exceeds 60 %, σ and E increase prominently with increasing plant powder content, reaching a maximum at plant particles/powder content = ca. 70 %, and then decrease with increasing plant powder content.

The DTG curves of the prepared PU composites show two kinds of thermal degradation temperatures: DT_{d1} and DT_{d2}. The DT_{d1} decreases with increasing plant powder content. The DT_{d2} increased slightly with increasing plant powder content.

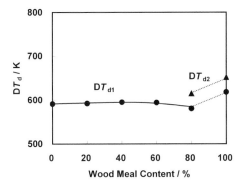

Figure 8-6. The change of DT_{d1} and DT_{d2} of wood meal-MP type PU composites with wood meal contents. ●: DT_{d1}, ▲: DT_{d2}

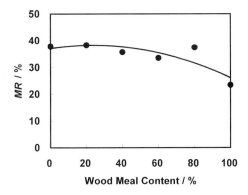

Figure 8-7. Change of mass residual amount (*MR*, %) at 723 K and wood meal content in PU composites.

2. BIOCOMPOSITES CONTAINING COFFEE GROUNDS

Polyurethane (PU) composites that are prepared from ground plant particles, such as coffee grounds, mixed with a molasses-polyol (MP) solution consisting of molasses and polyethylene glycol (PEG 200) are described in this section. Mechanical and thermal properties of the above composites are also considered.

2.1   Preparation

PU composites containing coffee grounds (CG) as fillers can be prepared according to the scheme shown in Figure 8-8 [7,8]. CG are first mixed with polyol containing molasses or lignin. The suspensions with various mixing ratios from 10, 20, 30, 40, 50, 60, 70, 80 and 90 wt % of CG in molasses polyol (MLP) are prepared. Lignin-based polyol such as kraft lignin-based polyol (KLP) can also be used. Acetone may be added to each mixture in order to control the viscosity of the suspension. MDI is added to the suspension under stirring and PU composites are prepared. After drying at room temperature, the sample is cured at 393 K for 2 hrs.

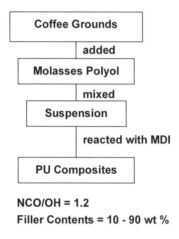

Figure 8-8. Preparation scheme of polyurethane composites (PU composites) containing coffee grounds (CG) in molasses polyol (MLP) [7].

2.2   Thermal and mechanical properties

Figure 8-9 shows the change of density (ρ) of PU composites with CG contents. The density reaches a maximum when CG content in MLP is ca. 70 %.

Figures 8-10 and 8-11 show the change of compression strength (σ) and modulus of elasticity (E) of PU composites with CG contents in MLP and KLP. As seen from the figure, compression strength (σ) and modulus of elasticity (E) increase with increasing CG contents in PU composites and reach a maximum when CG content is ca. 70 % in KLP type PU composites and ca. 80 % in MLP type PU composites.

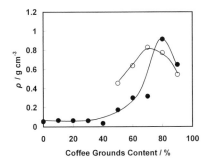

Figure 8-9. Change of density (ρ) with coffee grounds (CG) content in polyols such as MLP and KLP. ●: MLP type PU composites, ○:KLP type PU composites

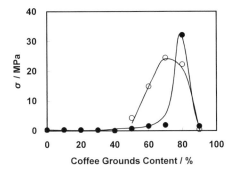

Figure 8-10. Change of compression strength (σ) with coffee grounds (CG) content in polyols such as MLP and KLP. ●: MLP type PU composites, ○:KLP type PU composites.

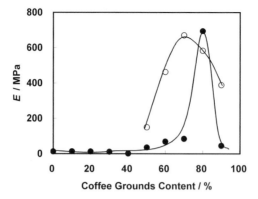

Figure 8-11. Change of modulus of elasticity *(E)* of PU composites with coffee grounds content. ● MLP type PU composites ○ KLP type PU composites.

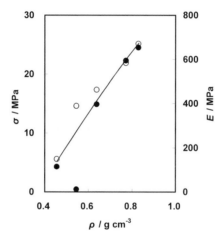

Figure 8-12. Change of compression strength (σ) and compression elasticity (*E*) of PU composites containing CG with apparent density (ρ). ●: σ, ○: *E*.

Figure 8-12 shows the change of compression strength (σ) and compression elasticity (*E*) of PU composites with apparent density (ρ). As clearly seen from the figure, σ and *E* increase almost linearly with increasing ρ, showing the strong dependency of mechanical properties of the PU composites on ρ.

Figure 8-13 shows TG and DTG curves of PU composites prepared from CG. As seen from Figures 8-14 and 8-15, TG and DTG curves show the presence of three kinds of thermal degradations corresponding to T_{d1}, T_{d2} and

T_{d3}, DT_{d1}, DT_{d2} and DT_{d3}. T_{d2}, T_{d3}, DT_{d2} and DT_{d3} seem to be specific to the degradation of CG, since those peaks are prominent when CG content is 100 %.

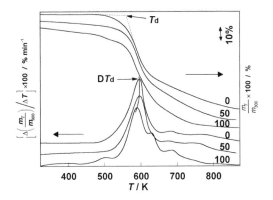

Figure 8-13. TG-DTG heating curves and derivative curves of MLP type PU composites containing various amounts of coffee grounds.

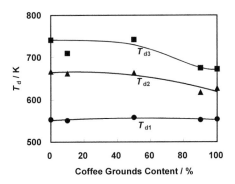

Figure 8-14. Change of T_d with coffee grounds content in MLP type PU composites.

Figure 8-16 shows the change of *MR* with CG contents in PU composites. The results show that CG parts in the composites degrade at 723 K, which is more easily than polyurethane parts of the composites, since thermal degradation proceeds more efficiently with increasing CG contents.

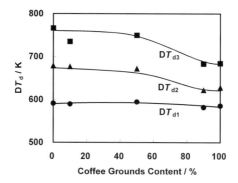

Figure 8-15. Change of DT$_d$ with coffee grounds content in MLP type PU composites.

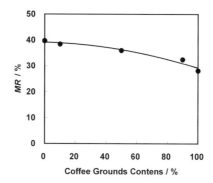

Figure 8-16. Change of mass residue *(MR)* at 723 K with coffee grounds content in MLP type PU composites.

3.   GEOCOMPOSITES

In two major components of plant materials such as cellulose and lignin, lignin is a promising biomass, which is obtained as a by-product of pulp and paper industries and has not been effectively utilized until now. Lignin is usually considered as a polyphenolic material having an amorphous structure, which arises from an enzyme-initiated dehydrogenative polymerization of coniferyl, sinapyl and *p*-coumaryl alcohols [12-14]. Therefore, the basic lignin structure is classified into two components; the aromatic part and the C3 chain part having propane-unit structure. The only usable reaction site in lignin is the OH group, which is the case for both phenolic and alcoholic hydroxyl groups. Molasses is also obtained as a by-product of the sugar industry, having alcoholic hydroxyl groups as the reactive site.

In the polyurethane (PU) preparation, the hydroxyl groups in plant components are effectively used as the reactive site. The PU's prepared from plant components are not only biodegradable but also show physical properties which can be satisfactorily used in practical fields in various industries such as construction and packaging. In this section, new types of PU geostabilizers derived from kraft lignin (KL), sodium lignosulfonate (LS) and molasses (ML) are described. Preparation of geocomposites which are prepared by the reaction of PU-based geostabilizers in sand and the mechanical and thermal properties of the above geocomposites are considered in this section.

3.1   Preparation

Three kinds of polyol were prepared; one portion of KL, LS or ML is dissolved in 2 portions of polyethylene glycol with average molecular mass 200 (PEG 200) or triethylene glycol (TEG). The above polyols are designated as KLP, LSP, MLP, KLTP, LSTP and MLTP. Polyols such as KLP, LSP and MLP are mixed with PEG 200 with various mixing ratios. Polyols such as KLT, LST and MLT are also mixed with TEG with various mixing ratios. The amount of KL, LS or ML is defined as follows.

KL, LS or ML content in polyol = [(mass of KL, LS or ML) /(mass of PEG 200 or TEG)] x 100, %   (8.1)

Table 8-1. Samples

Biomass	Abbreviation	Polyol	Symbol
Kraft Lignin (KL)	KLPPU	polyethylene glycol (PEG)	●
	KLTPU	triethylene glycol (TEG)	○
Lignosulfonate (LS)	LSPPU	PEG	▲
	LSTPU	TEG	△
Mollases (ML)	MLPPU	PEG	■
	MLTPU	TEG	□

KL, LS and ML contents in polyol are 0, 3.3, 6.6, 9.9, 16.5, 19.8, 23.1, 26.4 and 29.7 %, respectively. In order to prepare polyurethane geocomposites, MDI as an isocyanate, dibutyltindilaurate (DBTDL) as a catalyst, distilled water as a foaming agent and silicon surfactants as foam controlling agents are used [15, 16]. Six kinds of polyols are prepared as listed in Table 1 together with the abbreviations.

Test sample pieces of geocomposites of the KL series are prepared as follows [15,16]: (1) ca. 0.270 kg of silicate sand (Japanese Industrial Standard, JIS No 4) was dried in an oven controlled at 300K for 30 minutes, (2) dried silicate sand is filled in a polypropylene (PP) cylinder with diameter 4.0×10^{-2} m and length 2.1×10^{-1} m equipped with a lid coated with fluorine type removing agent, (3) 100 ml of water is added and an excess amount of water was excluded from the sand using a corking hand gun. In this stage, the mass of sand increases ca. 16 %, (4) the surface of the sand is flattened, (5) a pre-determined amount of PEG 200, a small amount of foaming agent, foaming controlling agent and DBTDL are added to pre-determined amount of KLP under stirring, (6) the mixture is stirred for 1 min and then MDI is added. NCO/OH ratio is adjusted to 1.4. The total amount of solution is 0.030 kg, (7) before drastic foaming starts, the solution is quickly poured into the sand, (8) an injection syringe equipped with an o-ring is inserted in the PP cylinder and the content was compressed using a corking hand gun, (9) the sand containing prepolymers stands for 24 hours under compression at 300K, (10) solidified sand is taken from the cylinder and non-reacted sand was removed. Geocomposites with LS or ML series are prepared in a similar manner as stated above.

3.2   Permeation distance

The samples containing MLPU are dark yellow, KLPU brown and LSPU dark brown. No sand comes off from the surface, although the top and side surfaces were smooth and the bottom face is uneven. Figure 8-17 shows a prepared sample.

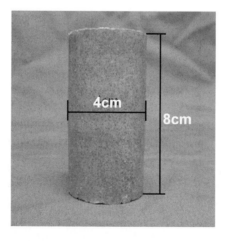

Figure 8-17. Photograph of prepared geocomposites. When the samples are taken out from the PP cylinder, the top and side surfaces of the samples are smooth but the bottom surface is

uneven. Using a digital caliper, the maximum length from the top to bottom (l_{max}) and the minimum length (l_{mim}) are measured, as shown in Figure 8-17. Permeation length of prepolymer in the sand was defined as the average of l_{max} and l_{mim}.

Figure 8-18 shows change of permeation distance of geocomposites as a function of KL, LS and ML contents in PEG and TEG solutions. Permeation distance is defined as shown in equation 8.2. Measurement method is found in the caption of Figure 8-17.

$$\text{Permeation distance} = \frac{\left(l_{max} + l_{min}\right)}{2} \tag{8.2}$$

Figure 8-18 shows change of permeation distance of LSTPU and LSPPU with LS contents in TEG and PEG solutions. Figure 8-18 shows change of permeation distance of geocomposites with KL, LS and ML contents in PEG and TEG solutions. Permeation distance of the above samples increases in initial stage by adding lignin. As shown in Figure 8-18, water insoluble KL shows quite different behaviour compared with water soluble LS and ML. Permeation distance of KLPPU increases in the initial state and reaches the maximum point at 10 %. After exceeding the maximum point, permeation distance of KLTPU and KLPPU markedly decreases. In contrast, permeation distance of LSPPU, LSTPU, MLPPU and MLTPU maintain an almost constant value after the initial increase. Permeation distance is affected by various factors, the major one is viscosity of injected solution. At the same time, chemical properties of lignin solved in the solution should be taken

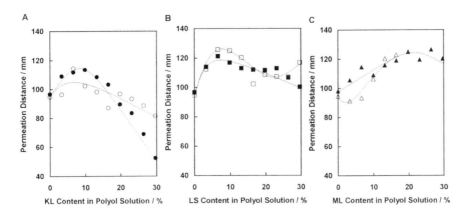

Figure 8-18. Change of permeation distance of geocomposites with KL, LS and ML contents in TEG and PEG. A, O: KLTPU, ●: KLPPU, □: B, LSTPU, ■: LSPPU, C, △: MLTPU, ▲: MLPPU.

into consideration. Hydrophylicity of LS is far larger than that of KL, since LS contains ionic sulfonate groups in their side chains. On this account, hydrophylicity of LS plays an important role, when solution is diffused in wet sand matrix.

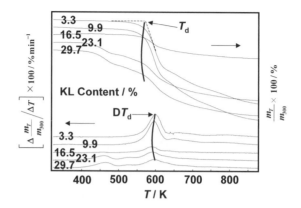

Figure 8-19. Stacked TG curves and derivatives TG curves of a series of KLTPU. Measurements: thermogravimeter (TG)-differential thermal analyzer (DTA) (Seiko Instruments TG/DTA220), heating rate = 10 K min^{-1}, sample mass =ca. 7 mg, N$_2$ flow rate= 100 ml min^{-1}. Decomposition temperature was defined as shown in Figure 2-3 (Chapter 2). Peak temperatures of derivative TG curves (DT_d) are also measured. The mass residues are obtained at 723, 773 and 823 K.

Figure 8-19 shows representative TG curves and their derivative TG curves (DTG) of KLTPU samples with various KL contents. The arrows indicate decomposition temperature (T_d) and peak temperature of derivative TG curves (DT_d). Compared with PU foam, heating rate dependency of composites is large due to heat conductivity. As reported previously [18], PU derived from KL is decomposed in two stages at around 560 and 620 K. The first decomposition is attributed to dissociation of isocyanate and phenolic hydroxyl groups of lignin and the second one is decomposition of PEG chain [18].

Similar TG curves and DTG curves were obtained for the other samples. Since the content of lignin in the composites is small, the 2nd step thermal decomposition is scarcely observed. On this account, the first step decomposition is examined in this study.

Figure 8-20 shows change of T_d with KL, LS or ML contents for KLPPU, LSPPU and MLPPU samples. As shown in Figure 8-20, T_d ranges from 530 to 550 K. T_d values of the samples decrease gradually with increasing KL, LS or ML contents. As shown in Figure 8-21, DT_d values of the above samples maintain an almost constant value. The above results suggest that the amount of lignin or molasses does not accelerate the decomposition. This tendency is reasonable, since the content of PU is small.

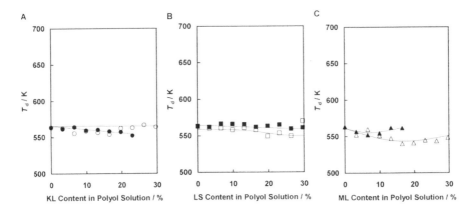

Figure 8-20. Decomposition temperature (T_d) and peak temperature of TG derivative curves (DT_d) of geocomposites as a function of KL, LS or ML content in PEG or TEG solution. A, KL, O: KLTPU, ●: KLPPU, B, LS, □: LSTPU, ■: LSPPU, C, ML, △: MLTPU, ▲: MLPPU.

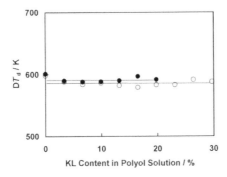

Figure 8-21. Peak temperature DT_d) of TG derivative curves of geocomposites as a function of KL content in TEG and PEG solutions. O: KLTPU, ●: KLPPU.

Figure 8-22 shows mass residue of KLTPU, KLPPU, LSTPU, LSPPU, MLTPU and MLPPU at 723 K as a function of lignin or molasses content in PEG. The amount of mass residue (*MR*) ranges from 50 to 60 % in KLTPU, LSTPU and MLTPU, and no large difference is observed for the three series of samples. The amount of *MR* ranges from 80 to 90 % in KLPPU, LSPPU and MLPPU. Even when mass residue at 773 K was calculated and compared with that of 723K, the difference was almost negligible. The major part of the residue is silicate sand, and the residual carbon which is obtained by thermal decomposition of polyurethane foams is less than 7 % in

the geocomposites. The mass residue of KLTPU, LSTPU and MLTPU
samples was smaller than that of KLPPU, LSPPU and MLPPU samples.

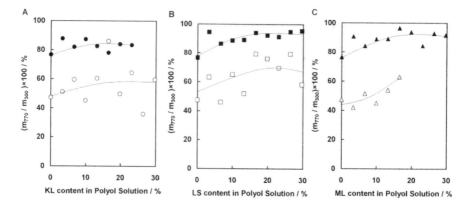

Figure 8-22. Change of mass residue (*MR*) at 723 K with KL, LS or ML content in PEG
solution. A, O: KLTPU, ●: KLPPU, B, □: LSTPU, ■: LSPPU, C, △: MLTPU, ▲: MLPPU.

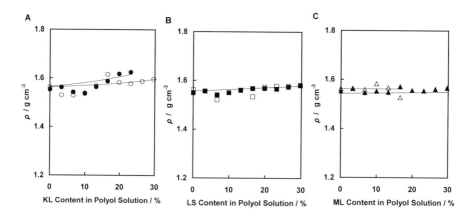

Figure 8-23. Apparent density of geocomposites as a function of KL, LS and ML content in
TEG and PEG solution. O: KLTPU, ●: KLPPU, G: LSTPU, ■: LSPPU, △: MLTPU, ▲:
MLPPU. Measurements; Sample pieces for mechanical tests with a diameter 4.0×10^{-2} m and
length 8.0×10^{-2} m were prepared using an automatic cutter. Average values of diameter and
length are measured at three different points and volume (*V*) is calculated. The mass (*m*) of
three samples is weighed using an electric balance EB-4300DVW. Apparent density is
calculated ($\rho = m/V$, g cm^{-3}) using the averaged values.

Figure 8-23 shows apparent density (ρ) of geocomposite samples as a
function of contents of lignin or molasses in polyol. As clearly seen, no large
variation is observed if the kind of polyol or amount of plant materials are
varied. The apparent densities of KLPPU increase slightly at around KL

content 16 to 25 %. This slight increase accords with the marked decrease of permeation distance of KL at around a similar KL content as shown in Figure 8-18. The apparent density of polyurethane foam without sand prepared under similar conditions is ca. 0.1 g cm-3. This suggests that PU foams are formed on the surface of the sand and the mass of PU is negligible for ρ of composites.

When compression strength (σ) of KLPPU, LSPPU and MLPPU is plotted against content of lignin or molasses in PEG, σ increases slightly in the initial stage and then decreases linearly with increasing lignin and molasses contents as shown in Figure 8-24. The curves decrease in a similar manner in two series of KL and LS with TEG and PEG. The σ values of MLPPU are slightly larger than those of lignin-based PU's. As shown in Figure 8-24, the maximum values of geocomposites with TPU are larger than those of the samples containing PPU and the decrease of σ is rapid compared with those of PPU series samples. The results shown in Figure 8-28 indicate that an optimum amount of lignin or molasses exists where the compression strength of geocomposites shows the highest value. It is generally accepted that the compression strength of geocomposites should be larger than 2 MPa. Accordingly, values of the samples using PEG are acceptable for practical applications. The above facts suggest that the chain length of ethylene glycol, i.e. molecular mobility of soft segments, is an important factor. Compression modulus varies in a similar manner to compression strength as shown in Figure 8-25.

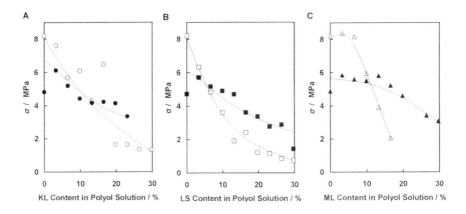

Figure 8-24. Change of compression strength (σ) of geocomposites at 10% with KL, LS or ML contents in TEG and PEG solution. O: KLTPU, ●: KLPPU, □: LSPPU, ■: LSTPU, △: MLTPU, ▲: MLPPU. Measurements; A Shimadzu Autograph AG-2000D is used for uni-axial compression strength tests. The diameter of the sample was 4.0×10^{-2} m and the length is 8.0×10^{-3} m. Compression rate is 1.0×10^{-4} m min^{-1} Compression stress (σ, Pa) is calculated according to JIS A1216.

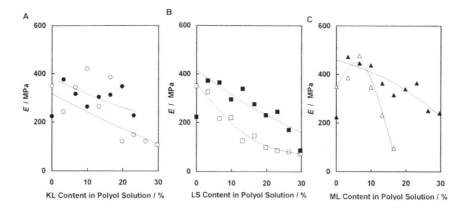

Figure 8-25. Change of compression modulus (*E*) of geocomposites with KL, LS and ML contents in TEG or PEG solution. ○: KLTPU, ●: KLPPU, □: LSTPU, ■: LSPPU, △: MLTPU, ▲: MLPPU.

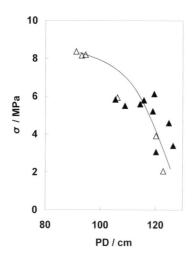

Figure 8-26. Change of σ with permeation distance of MLTPU and MLPPU. △: MLTPU, ▲ : MLPPU.

As discussed above, the total amount of PU in composites necessarily relates to the mechanical properties. On this account, the relationship between σ and permeation distance is established. Figure 8-26 shows the relationship between compression strength and permeation distance of MLTPU and MLPPU. Compression strength values decrease with increasing permeation distance for both MLTPU and MLPPU. It is reasonable to consider that σ values decrease when the amount of ML increases. Concerning the samples containing lignin, similar relationships are established. The above experiments suggest that compression strength is markedly improved in the presence of a small amount of PU in the sand.

REFERENCES

1. Hatakeyama, H., Hirose, S., Nakamura, K. and Hatakeyama, T., *New types of Polyurethanes Derived from Lignocellulose and Saccharides,* in *Cellulosics: Chemical, Biochemical and Material Aspects*, J. F. Kennedy, G. O. Phillips and P. A. Williams, Eds., Ellis Horwood, 1993, pp. 525-536
2. Hirose, S., Yano. S., Hatakeyama, H. and Nakamura, K. , 2000, *JPat*. 3101701.
3. Hirose, S., Yano. S., Hatakeyama, H. and Nakamura, K. , 1991, *USPat*. 4987213.
4. Hirose, S., Yano. S., Hatakeyama, H. and Nakamura, K. , 1989, *DEPat*. 3884781.
5. Hirose, S., Yano. S., Hatakeyama, H. and Nakamura, K. , 1990, *BRPat*. 8902248.
6. Hatakeyama, H., Hirose, S., Kobashigawa, K. and Tokashiki, T., 1998, JPat. 2836004.
7. Hatakeyama, H., Hirose, S., Nakamura, K. and Kobashigawa, K., 1997, JPat2613834.
8. Hatakeyama, H., Hirose, S., and Nakamura, K., 1997, JPat2611171.
9. Hatakeyama, H., 2002, Thermal analysis of environmentally compatible polymers containing plant components in the main chain. *J. Therm. Anal. Cal.* **70**, 755-759.
10. Hatakeyama, H., Asano, Y., and Hatakeyama, T., 2003, Biobased polymeric materials. In *Biodegradable Polymers and Plastics* (Chellini, E. and Solario, R. eds.), Kluwer Academic / Plenum Publishers, New York, pp. 103-119.
11. Morohoshi, N., Hirose, S., Hatakeyama, H., Tokashiki, T. and Teruya, K., 1995,Biodegradability of polyurethane foams derived from molasses. *Sen-i Gakkaishi*, **51**, 143 –149.
12. Sarkanen, K. V. and Ludwig, C. H., 1971, *Lignins, Occurrence, Formation, Structure and Reactions.* New York: John Wiley & Sons, Inc.
13. Goring, D. A. I., 1989, *The lignin paradigm. In: Lignin, Properties and Materials* (Glasse, W. G and Sarkanen S. eds.), Washington, DC: ACS Sym. Ser., 397, Amer. Chem. Soc., p p. 2-10.
14. Lin, S. Y. and Dence, C. W., 1992, Methods *in Lignin Chemi*stry. Heidelberg, Springer-Verlag.
15. Wakisaka, O., Mizuno, N., Ikaga, S., Hatakeyama, H. and Hirose, S., 2002, JPat 2002146351.
16. Wakisaka, O., Mizuno, N., Ikaga, S., Hatakeyama, H. and Hirose, S., 2002, DE10153980.
17. Hatakeyama T. and Quinn, F. X., 1999, *Thermal Analysis*. Chichester, John Wiley & Sons.
18. Hirose, S., Kobashigawa, K., Izuta, Y, and Hatakeyama, H., 1998, Thermal degradation of polyurethanes containing lignin studied by TG-FTIR. *Polym. Inter.*, **47,** 247-256.

Index

180-τ-90 degree pulse method 102, 206
^{1}H longitudinal relaxation time
   amorphous cellulose 84
   NaCS-water 102, 102
   NaLS-water 205
^{1}H transverse relaxation time
   amorphous cellulose 52
   NaCS-water 101, 102
   lignin 206

^{23}N transverse relaxation time
   NaCS-water 106
   NaLS-water 207
^{23}Na longitudinal relaxation time
   NaCS-water 105
   NaLS 207

activation energy
   b-NMR 194, 206
   CAPCL 224
   DMA 224
   DSC 52
   NaCS-water 107
   TG 121
AFM 3, 4, 5, 6
alginic acid 2, 132
ALPCL
   MR 242
   T_d 242
   T_g 241

T_m 241
ALPCL PU 291
apparent density 26
   KLDPU 281
   KL-ML-PEG200-MDI 302
   KLPPU 281
   KLTPU 281
   LS-ML-PEG200-MDI 302
   PEP-GP-MLP-(LDI/LTI) 260

baseline optical density 32
b-NMR
   line shape 191
   line width 191
bound water
   cellulose 65
   crystallinity 68
   dioxane lignin 202
   lignin 202
   natual cellulose 67
   regenarated cellulose 67
breaking strength
   cellulose 63
   cellulose-water 64
Brunauer-Emmett-Teller equation 28
   BET constant 199

CAPCL 218
   DMA 223
   melting enthalpy221
cellopentaose 55

cellotriose 55
CellPCL 230
    DSC curve 231
    glass transition 232
    melting 233
    TG 233, 234
    TG-FTIR 235, 234, 236
cellulose 1, 39
     amorphous cellulose 43, 49, 51, 53,
    78, 83
    cellulose acetate 41, 45
    crystalline structure 40
    crystallinity 43
    D$_2$O 79
    heat capacity 43, 65
    hollow fibre 76
    psuedo gel 108
    regenerate cellulose-water 60
    stress-strain 63
    viscoelasticity 72
    water-cellulose 60
    WAX 61
CL/OH ratio
    CAPCL 221, 226
    CellPCL 232
compression elasticity
    KLDPU 283
    KLPTPU 283
    KLTPU 283
compression strength
    KLDPU 281
    KL-ML-PEG200-MDI 301
    KLPPU 281
    KLTPU 281
    LSDPU 289
    LS-ML-PEG200-MDI 301
    LSPPU 289
    LSTPU 289
    PEP-GP-MLP-(LDI/LTI) 260
    PEP-PPG-MLP-(TDI/MDI) 273
    PEP-PPG-MLP-MDI 272
compression test 25
correlation time
    lignin 205
    NaCS-water 104
cotton fabric
    life time 123
creep 20
crosslinking density 250

curdlan 148

DBTDL 261, 275
degree of substitution
    cellulose acetate 44
    CMC-cations 87
    NaCMC 47
    non-freezing water 89
    pectin 154
deueration
    IR sample holder 33
    amorphous cellulose 80, 82
differential thermal analysis 13
diffusion constant
    amorphous cellulose 82
DMA 13, 23
    apparatus 24
    CAPCL
    humidity 24
DPPH 198
DSC 13
    heat-flux type DSC 17
    power compensation type DSC 18
DSC curve
    LS-ML-PEG200-MDI 297
    ALPCL 239
    ALPCLPU 291
    amorphous cellulose 51
    amorphous cellulose-water 78
    CaAlg-water 138
    CAPCL 220
    CellPCL 231
    cellulose acetate 45
    cellulose gel 109,110, 111
    CMC-cations-water 87
    curdlan 149
    fractionated lignin 180
    guar gum (GG)-water 164
    hollow fibre 77
    KL-ML-PEG200-MDI 296
    KLDPU 277
    KLPCL 240
    KLPPU 277
    KLTPU 277
    lignin 175, 181
    lignosulfonate 204
    LSDPU 285
    LSTPU 285
    methylcellulose 114

NaAlg-water 133
NaCMC-water 85
NaCS-water 94
natural cellulose 59
PEP-GP-MLP-(LDI/LTI) 255
PEP-PPG-MLP-(MDI/TDI) 264
poly(4-hydroxy,3-methoxystyrene)
   187
poly(4-hydroxystyrene) 187
polystyrene 187
regenerated cellulose 59
saccharides 54
schematic 19
vaporization 69
water 58, 66
xanthan gum-water 156, 159

dynamic modulus
   CAPCL 223
   celllophane-water 73
   cellophane 72
   xanthan gum 144

electron spin resonance spectrometry 33
enthalpy relaxation 185
ESR 33
   spectra of MWL 197
   radical 198

falling ball method 30, 142
freezing and thawing 150
   LBG 151
fructose 7, 251

gel
   alginic acid 132
   CaAlg 139
   cellulose gel 108, 111, 112
   CMCPU hydrogel 114
   curdlan 148, 150
   gel sol transition, 150, 142
   hydrogel 108
   LBG 152
   pectin 153
   xanthan gum 142
gel permeation chromatogram
   lignin 179, 180
glass transition 174
   ALPCL 240

ALPCL PU 291
CAPCL 219
celllose acetate 44
CMCPU hydrogel 115
CellPCL 232
fractionated lignin 181
KLPCL 240
KLPCL PU 291
KLDPU 278
KL-ML-PEG200-MDI 297
KLPPU 278
KLTPU 278
lignin 175
LSDPU 286
LSPPU 286
LSTPU 286
NaCMC 48
NaCMC-water 92
PEP-GP-MLP-(LDT/LTI) 256
PEP-PPG-MLP-(MDI/TDI) 265
PEP-PPG-MLP-MDI 263
polystyrene 178
PPG-MLP-MDI 263
glucose 7, 251
guar gum 159
guluronic acid 132

heat capacity
   amorphous cellulose -water 83
   annealed dioxane liginin 185
   cellulose 42, 43
   cellulose-water 64
   CMC-water 91
   dioxane lignin 184
   NaCMC 47, 48
   NaCMC-water 92
   NaCS-water 97
   saccharides 56
hollow fibre
   cellulose 73
   cellulose triacetate 75, 76
   DSC 75
   pore size distribution 77
   SEM 75
   spinning apparatus 74
   water 75

ionic radius
    Alg 137
IR peak intensity
    CAPCL 228, 229, 230
    CellPCL 236, 237, 238
    KLPCL 244, 246
    PEP-GP-MLP-(LDI/LTI) 257
IR spectra
    amorphous cellulose 80
    CellPCL 235
    lignin 176
KLDPU 276
KL-ML-PEG200-MDI 295
KLPCL 238
    MR 242
    T_d 242
    T_g 241
    T_m 241
KLPPU 276
KLTPU 276
kraft lignin-based PCL → KLPCL

life time prediction 122
Lig PCLPU 290
lignin 2, 7, 171
    basic structure 172
    b-NMR 192, 193
    Björkman lignin 176
    dioxane lignin 175, 177
    DSC heating curve 181
    geocomposite 314
    glass transition 173
    kraft lignin 175, 176, 177
    lingnosulfate 177, 193, 289, 293
    lignin-based PCL 238
    lignin-based PU 275, 283, 293
    methylated dioxane lignin 182
    milled wood lingin 177
    MWL 173, 175
    NMR relaxation 203
    water interaction 200202
    WAX 173
linear expansion coefficient 20
liquid crystal
    enthalpy 99
    NaCS 98
    transition enthalpy 107
    WAXS 101
    xanthan gum 158

Lissajous diagram 22
local mode relaxation 23
    CAPCL 223
    cellulose-water 73
    lignin 196
locust bean gum 2, 150, 159
LSDPU 285
LS-ML-PEG200-MDI 295
LSPPU 286
LSTPU 286
lysine diisocyanate 253

main chain motion
    b-NMR 196
    CAPCL 223
    DMA 23
    lignin 189, 196
manuronic acid 132
mass residue
    CAPCL 227
    CellPCL 234
    geocompoiste 320
    KLDPU 280
    KLPPU 280
    KLTPU 280
    LSDPU 288
    LSPPU 288
    LSTPU 288
    MLP type PU 314
    PEP-GP-MLP-(LDI/LTI) 259
    PEP-PPG-MLP-(MDI/TDI) 271
    PPG-MLP-MDI 270
    wood meal type PU 309
MDI 274
Meiboom-Gill Carr-Purcell method
    NaCS-water 102
    NaLS 206
methylcellulose 42, 113
molasses 7
    polyol 252
molecular mass
    lignin 178
    NaCMC 47
molecular mass distribution
    lignin 178
    MWL 173, 176, 177

NaAlg
    phase transition 133

NaCMC
    AFM 4
    cation 87
    glass transition 48,49
    heat capacity 48, 92
    non-freezing water 88, 90
    phase diagram 86
    hydrogel 114
    water-NaCMC 85
    substitution 89

NaCS 5
    AFM 5,6
    heat capacity 97
    liquid crystal 98
    NMR relaxation 101
    phase transition 93

NaLS 6, 203.
non-freezing water
    Alg 137
    celluloseI 67, 68
    cellulose gel 112
    C_p 91
    dioxane lignin 203
    guar gum 162
    ionic radius 90
    LBG 152, 162
    MgCMC 90
    NaAlg 135
    NaCMC 88, 89, 90
    NaCS 96
    vaproization 70
    tara gum 162
    xanthan gum 157

Ozawa-Wall-Flynn method 119, 210
    lignin 210

PCL
    ALPCL 239
    cellulose-based 230
    cellulose acetate-based 219
    KLPCL 240
    lignin-based 238
pectic acid 153
pectin 152
PE-GP-MLP- (LDI/LTI) 255
PEP-PPG-MLP-(MDI/TDI) 268, 271

PEP-PPG-MLP-(TDI/MDI) 273
PEP-PPG-MLP-MDI 267, 270, 272
permeation distance 316
phase diagram
    ALPCLPU 292
    cellulose-water 60
    CMCPU hydrogel 116
    guar gum-water 162
    LBG-water 161
    NaAlg-water 134
    NaCMC-water 86
    NaCS-water 95
    NaLS-water 204
    tara gum-water 161
    xanthan gum-water 157
polarizing light micrograph
    NaCS 98
    CaAlg fibre 139
    xanthan gum 158
polarizing light microscopy 31
poly(4-hydroxy, 3,5-methoxystyrene) 186
poly(4-hydroxy, 3-methoxystyrene) 186
poly(4-hydroxystyrene) 186
poly(vinyl alcohol) 150
poly(vinylpyrolidone) 76
polymorphism
    cellulose 41
polystyrene 173, 178
polyurethane $\rightarrow$ PU
pore size
    hollow fibre 76
PPG-MLP-MDI 262, 265, 266
PU
    CMC hydrogel 115
    flexible foam 253
    KL-based 276, 294
    lignin-based 273
    LS-based 284, 294
    ML-based 294
    rigid foam 294
    semi-rigig foam 261
    sucrose-based 250
PVP 76, 78

quadrapole relaxation 206

radical
    lignin 197
relative humidity 199

viscoelasticity 71
relative optical baseline density
   amorphous cellulose 82
   lignin 176

SAXS
   xanthan gum 144, 145
second moment
   b-NMR 194, 196
solid echo method
   NaCS-water 102
spin probe method 197
stress relaxation 20
stress-strain curve 25
   cellulose-water 63
   KLDPU 280
   LSDPU 289
   LSPPU 289
   LSTPU 289
   schematic 26
sucrose 7, 251
swelling
   NaAlg 139
   swelling ratio 29
synchrotron orbital radiation 144

tara gum 159
TBA
   lignin 188, 189, 190
   relative humidity 201 TEG 283
tensile property
   DL 191
   KL 191
TG, 13, 14
TG, DTG curve
   CAPCL 225
   CellPCL 233
   cellulose 121, 122
   dioxane lignin 211
   glucose 117
   KLDPU 279
   KL-ML-PEG200-MDI 298
   KLPCL 242
   KLTPU 318
   LSDPU 287
   LS-ML-PEG200-MDI 299
   MLP type PU 313
   PEP-GP-MLP-(LDI/LTI) 256
   PEP-PPG-MLP-(MDI/TDI) 269

PEP-PPG-MLP-MDI 267
PPG-MLP-MDI 265
schematic 15
solvolysis lignin 208
wood meal-MLP type PU 308
TG-DTA 13
TGFTIR 13
   CAPCL 227
   CellPCL 235
   D-glucose 118
   KLPCL 243
   kraft lignin 209
thermal decomposition
   kinetics 119
   lignin 208
thermal decomposition temperature
   CellPCL 234
   glucose 118
   KLDPU 279
   KLTPU 319
   KL-ML-PEG200-MDI 300
   LSDPU 287
   LS-ML-PEG200-MDI 300
   LSPPU 287, 319
   LSTPU 287, 319
   MLPPU 319
   MLTPU 319
   PEP-GP-MLP-(LDI/LRI) 257
   saccharide 116

thermal history
   lignin 181
   NaCMC-water 85
thermogravimetry → TG
TMA 13, 19, 20
   dynamic measurement 22
   dynamic modulus 144, 155
   pectin 154
   probe 21
   swelling 22
   swelling curve 140
   compression mode 21
TMAEP 261
torsional braid analysis → TBA
transition enthalpy
   amorphous cellulose 52
   curdlan gel 149
   guar gum-water 165
   ligPCL 241

NaCS-water 99
xanthan gum-water 158

vaporization
cellulose-water 68
isothermal 70
non-freezing water 70
viscoelasticity 23
viscometry 30

water content
definition 57
water sorption 27, 199
WAX 33
amorphous cellulose 50, 79
cellulose-water 62, 61
lignin 173
NaCS-water 100, 101
wide angle x-ray diffractometry → WAX
xanthan gum 142, 155

yielding strength

KLDPU 282
KL-ML-PEG200-MDI 303
KLPPU 282
KLTPU 282
LS-ML-PEG200-MDI 303

ΔC_p
CAPCL 222
CellPCL 232
NaAlg-water 135
poly(4-hydroxystyrene) 188

α-dispersion
CAPCL 223

β-dispersion
CAPCL 224
lignin 196

ε-caprolactone → CL